Understanding Microbes

Understanding Microbes

An Introduction to a Small World

Jeremy W. Dale
University of Surrey, UK

⊛ **WILEY-BLACKWELL**

A John Wiley & Sons, Ltd., Publication

Registered office: John Wiley & Sons, Ltd, The Atrium, Southern Gate, Chichester, West Sussex, PO19 8SQ, UK

Editorial offices: 9600 Garsington Road, Oxford, OX4 2DQ, UK
The Atrium, Southern Gate, Chichester, West Sussex, PO19 8SQ, UK
111 River Street, Hoboken, NJ 07030-5774, USA

For details of our global editorial offices, for customer services and for information about how to apply for permission to reuse the copyright material in this book please see our website at www.wiley.com/wiley-blackwell.

Library of Congress Cataloging-in-Publication Data

Dale, Jeremy (Jeremy W.)
   Understanding microbes : an introduction to a small world / Jeremy Dale.
      pages cm
   Includes bibliographical references and index.
   ISBN 978-1-119-97880-0 (hardback) – ISBN 978-1-119-97879-4 (paper)   1. Microorganisms–
Textbooks.   2. Microbiology–Textbooks.   I.   Title.
   QR41.2.D35 2013
   579–dc23

                                                                          2012032317

A catalogue record for this book is available from the British Library.

Wiley also publishes its books in a variety of electronic formats. Some content that appears in print may not be available in electronic books.

Set in 10.5/13pt Times-RomanA by Thomson Digital, Noida, India

First Impression   2013

Contents

Preface

I suspect that you already know more about microbes than you think you do. You will have heard about microbes that cause disease, and you probably know that yeast is used in making bread, wine and beer, and that yoghurt contains 'friendly bacteria'. You may have noticed lichens on stones or the bark of trees; and even familiar fungi such as mushrooms can be considered as a form of microbe. However, the subject of microbes goes far beyond that. You may not know that there are ten times more bacteria in your intestines than there are human cells in your body, nor that microbes make up 50 per cent of the world's biomass and carry out as much photosynthesis as all the land plants combined. You may not be aware that, in the depths of the oceans, there are microbes that will only grow at temperatures above the normal boiling point of water, nor that up to 20 per cent of cancers are caused by viruses, nor that some ants cultivate fungal gardens to digest the plant material they bring back to their nest. Nor, indeed, that for the first two billion years of life on Earth, there were only microbes.

The purpose of this book is to introduce you to this 'small world' – a world which, although we don't realize it, is dominated by microbes. They were here long before we were, and they will be here after we have gone. Climate change may succeed in eliminating many of the forms of life we see around us, but the bacteria will survive.

Clearly, in a book this size, I cannot cover all of these aspects of microbial activity thoroughly. So, rather than reducing the subject to a set of lists, I decided to adopt a more selective approach. Some topics are dealt with at length, where I felt there was an interesting story to tell – and, conversely, a lot of topics are left out altogether, or dealt with rather briefly. Thus, this is not a systematic textbook, and I am sure that my colleagues will grumble about the omissions.

At the same time, to try to keep it accessible to a wider readership, I have tried to cut out unnecessary jargon and technical detail. It does get a bit technical in the molecular biology section (as a molecular biologist, I couldn't resist the temptation!) – but with the justification that these techniques have told us so much about what goes on that some understanding of them is needed. For example, you have to know a bit about genome sequencing to appreciate how we know that over 99 per cent of the microbes in the ocean were previously totally unknown. I hope you enjoy the result and that you come away feeling you want to find out more.

Finally, I owe a great debt of gratitude to my wife Angela, not only for tolerating me shutting myself away for days on end, but also for nobly reading right through a draft version, from the viewpoint of a non-microbiologist, and making many valuable suggestions. In the traditional manner, I have to absolve her from responsibility for any errors there might be, for which I have to be answerable.

Jeremy W. Dale

1
The Background

1.1 Meet the cast

The main point about microbes is that they are very small. Their one unifying feature is that they are too small to be seen without the aid of a microscope – although even that definition, as we will see later on, is blurred at the edges, as some of the 'microbes' I will consider are actually quite big. Within the basic definition, there is a substantial range of diversity. You will probably have heard of some of these microbes, such as the influenza virus or the bacterium known as MRSA, as they regularly make the news because they inflict themselves on us. Others may also be familiar because of their everyday role in fermentation – think of the yeasts that are needed for the production of bread, wine and beer, and the 'friendly' bacteria that are used for yogurt making. As we will see later on, there are very many more examples of a wide range of microbes that are of direct importance to us, both in disease (and health) and economically.

But this is only the tip of an enormous iceberg. Microbes are all around us, in vast numbers and diversity, especially in soil and water. It has been estimated that the world has 10^{31} bacteria – that's 1 with 31 zeros after it – with a total biomass greater than all the plants and animals combined (and that is just the bacteria, before we add in the other microbes – viruses, fungi, algae and protozoa). They play a massive role in shaping our environment, including fixing carbon and nitrogen from the air – and, by degrading organic matter, in releasing these elements again into the air. Our knowledge of the diversity of microbes in the environment has increased enormously in recent years. Molecular techniques that we will encounter in a later chapter have shown that most (perhaps 99 per cent) of these organisms were previously totally unknown and have never been grown in the laboratory.

We can begin the story in the 17th century, in Holland. Antonie van Leeuwenhoek was born in Delft, in 1632. After an apprenticeship with a cloth merchant, he set up his own drapery business and became a prosperous and influential citizen of Delft.

Having seen the magnifying glasses used by textile merchants for examining the cloth, he developed an interest in the use of lenses and started to make his own as a hobby. Although his 'microscopes' were simple by modern standards – consisting of just a single lens – the superb quality of his lenses, and his skill and patience in using them, enabled him to make many important observations. These included the first descriptions of microscopic single-celled organisms (which he called 'animalcules'), which he reported to the Royal Society in London in 1676.

These observations met with a considerable amount of scepticism but, eventually, after much further investigation, his achievements were recognized and he became a Fellow of the Royal Society in 1680. He continued to make many further detailed observations, such as the description of bacteria in plaque from teeth, until shortly before his death in 1723. Unfortunately, he kept secret some of the crucial details as to how he made his lenses, so, with his death, that part of the story came to an end.

However, others had also developed and used microscopes at around that time. Robert Hooke (1635–1703) is perhaps best known today for his study of the elasticity of materials, described by the mathematical relationship we still know as Hooke's Law. But that was far from the total of his interests or achievements, which ranged from experimental science to architecture. The part of his work we are concerned with here was his role in the development of the compound microscope (which, like a modern microscope, contained two lenses rather than the single lens used by van Leeuwenhoek).

He used this microscope to make a large series of observations of diverse biological materials, which was published in 1665 as a book, *Micrographia*. Notably, his description of the microscopic structure of slices of cork was the first identification of the cellular structure of, in this case, plant material (he coined the word 'cell' for them because of their resemblance to cells in a monastery). Although the microscopes used by Hooke (and other similar ones of that period) were more like a microscope of today than van Leeuwenhoek's single lens instruments, the technical difficulty of making them, and the superb craftsmanship of van Leeuwenhoek, meant that they were actually inferior to van Leeuwenhoek's.

It's now time we met the cast so, for the first members, let's consider viruses. These are so different from other microbes that it is only a matter of convenience that we do include them as 'microbes'. Indeed, it is questionable as to whether we should consider them as 'living' at all (I'll come back to that question in Chapter 10).

Viruses are not able to replicate, or to do anything at all, outside a host cell. The simplest viruses consist just of a nucleic acid molecule (which can be RNA or DNA, but not both), surrounded by a protein coat. These contain a limited number of genes. For example, one of the most basic viruses, called MS2, which infects *E. coli*, has just three genes: one codes for the coat protein, one is needed for copying the viral genome (RNA in this case) and the third is used to organize the assembly of the virus particle.

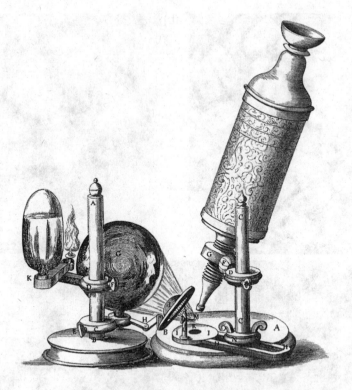

Figure 1.1 Robert Hooke's microscope. Drawing from 'Micrographia'.

However virus structures, and sizes, are quite diverse. Many, including some important human pathogens, are much larger than MS2 and have complex structures including, in some cases, a lipid coat. But virtually all viruses are so small that the human eye cannot see them, even with the aid of a light microscope –an electron microscope is needed to 'see' them. There are, however, some that are larger. The largest known viruses, such as the mimivirus which infects protozoa, are similar in size to some of the smallest bacteria, and they have a genome size to match. But even the largest and most complex viruses are completely unable to replicate without infecting a suitable host cell.

At several points within this book we will consider viruses that infect bacterial cells. These are called *bacteriophages*, or just *phages* for short (the word 'phage' being derived from the Greek '*phagein*', meaning 'to eat'). These are very wide-spread in nature and they can be of real practical significance – a phage infecting a bacterium that is used, for example, in the production of yoghurt can cause serious economic losses. But they have also played a major role in scientific research. Much of our knowledge of how genes work comes originally from studies of phages, where the simplicity of their structure, and the ease with which they can be grown and manipulated, made them invaluable models.

(a)   (b)

(c)

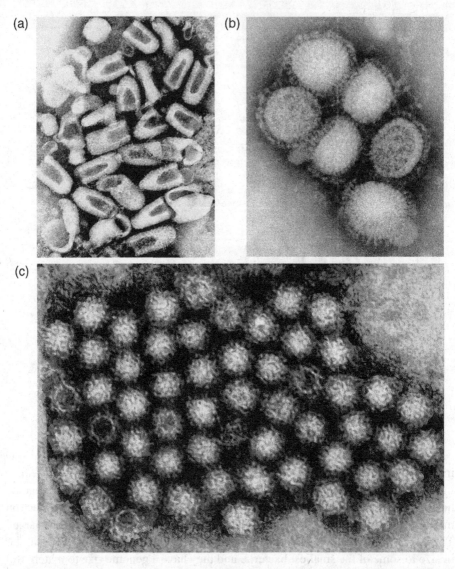

Figure 1.2   Electron micrographs of virus particles. (a) Rabies virus (courtesy of Frederick A Murphy http://www.utmb.edu/virusimages/); (b) influenza virus (colourized) (content provider(s): CDC/C. S. Goldsmith and A. Balish); (c) norovirus (colourized) (content provider(s): CDC/Charles D. Humphrey).

When a phage infects a bacterial cell, its nucleic acid (DNA or RNA, depending on the phage) is injected into the cell. Some of its genes are then recognized by the cell's machinery, which obligingly makes the relevant proteins that those genes code for. These proteins then divert the cell's activity away from its own genes and towards the production of many copies of the phage nucleic acid. At some point in this process, the DNA of the host cell is usually broken down and the bits are used for making the nucleic acid of the virus. The proteins that make up the external structure of the phage (the coat) are then produced, and the nucleic acid is packed into that structure. The consequence of this is the lysis of the bacterial cell and the liberation of hundreds or thousands of copies of the virus. The whole process, from infection to lysis may take perhaps 20–50 minutes (depending on the phage). The details of this process vary considerably from one phage to another, but the general principles are similar. An equivalent, but more complex, process occurs when viruses infect higher (eukaryotic) cells, including human cells.

Although we usually think of viruses as causing diseases, this does not always happen. Some viruses have the ability to remain latent within an infected cell. We may only realize that they are there when the latency breaks down, perhaps due to a drop in our immune defences. This happens, for example, with the herpes virus that causes cold sores, typically around the lips, and the varicella-zoster virus, which initially causes chickenpox but can subsequently remain dormant until causing an outbreak of shingles many years later.

The best understood example of latency is a virus (bacteriophage), known as lambda, that infects *E. coli*. This has a very elegant mechanism for ensuring that, in a proportion of newly infected cells, the expression of all the virus genes is turned off, apart from one gene that codes for a protein that is responsible for maintaining this repression. In this state, known as lysogeny, the DNA of the virus is integrated into the chromosome of the host cell and is therefore copied, along with the rest of the DNA, as the bacterial chromosome is copied during growth. Each daughter cell therefore receives a copy of the virus DNA. Studies of genome sequences have revealed that most bacteria (and, indeed, animal cells, including our own) contain a number of copies of a variety of viruses, stably integrated into the chromosome and never showing any signs of their presence.

Our second class of microbes is the bacteria and, by way of introduction, I will look at one bacterium in particular: *Escherichia coli*, or *E. coli* for short (unfortunately, most bacteria – and many other microbes – do not have simple common names, so we have to get used to using Latin ones). This is described as a rod-shaped organism, but it is better to visualize it as a short cylinder with rounded ends. Later on we will encounter bacteria with other shapes, especially ones such as *Staphylococcus*, which are spherical, as well as bacteria which grow as filaments or in spiral shapes.

E. coli, which is a common inhabitant of the human gut (as well as being able to cause some nasty diseases), is 2–3 μm long (a μm, or micrometre, is a millionth of a metre, or a thousandth of a millimetre) and 0.5–1 μm wide. It is a favourite model organism for bacteriologists because it will grow readily in a simple medium – all it needs is a sugar such as glucose and a nitrogen source such as an ammonium salt. It will grow even better if it is given a richer medium containing, for example, yeast extract. In a rich medium, it will divide every 20 minutes or so (bacteria typically grow in an apparently simple way – a cell gets bigger until it reaches a certain size, then it divides into two cells, which in turn grow and then divide again – so they multiply by dividing!).

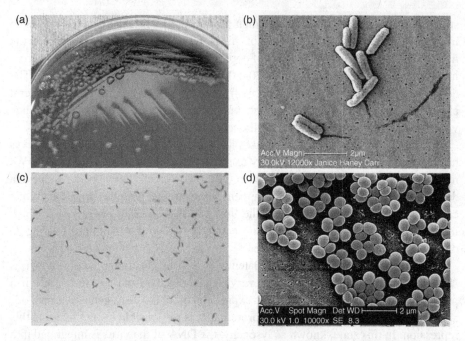

Figure 1.3   Bacteria. (a) Bacterial colonies on an agar plate (content provider(s): CDC/Amanda Moore, MT; Todd Parker, PhD; Audra Marsh); (b) Colourized electron micrograph of *Legionella pneumophila* (content provider(s): CDC/Margaret Williams, PhD; Claressa Lucas, PhD; Tatiana Travis, BS; Photo Credit: Janice Haney Carr); (c) Light microscopy of stained *Campylobacter* (content provider(s): CDC); (d) Colourized electron micrograph of *Staphylococcus aureus* (content provider(s): CDC/Matthew J. Arduino, DRPH; Photo Credit: Janice Haney Carr).

If we start with one cell, after 20 minutes there will be two, after 40 minutes four cells and, by one hour, eight cells. This is known as exponential, or logarithmic growth – it starts off slowly, but very soon reaches astronomical numbers. After ten divisions, there will be about one thousand cells; after another ten divisions, the number will reach a million; after a further ten divisions, it will be up to a thousand million cells.

When we get to such large numbers, they become very difficult to handle in the usual way, so we use what is known as scientific notation (see Appendix 2 for further explanation). A thousand, for example, is $(10 \times 10 \times 10)$ so we call it 10^3 rather than 1,000. A million (1,000,000) is 10^6, and a thousand million (1,000,000,000) is 10^9. So, after 30 cell divisions (about ten hours), we would have some 10^9 bacteria in our flask. This process does not continue indefinitely of course. After a while, the bacteria start to run out of nutrients (diffusion of oxygen into the medium is usually the first limiting factor for *E. coli*) and they will stop growing. For *E. coli*, this will usually happen when the concentration of bacteria reaches about 10^9 cells per millilitre. In other words, a 5 ml teaspoon would contain five billion, or five thousand million, bacteria.

One practical consequence of these massive numbers is worth a slight digression here. Disinfectant manufacturers will commonly make claims such as 'kills

99 per cent of household germs'. This sounds impressive – until we consider the numbers involved. If we start with say 10^6 bacteria (which is not really very high), then killing 90 per cent (or leaving ten per cent remaining) will reduce the numbers to 10^5; even killing 90 per cent of those (which leaves one per cent of the original), there will still be 10^4 bacteria. So killing 99 per cent (or leaving one per cent untouched) merely reduces the numbers from 10^6 to 10^4 (which we refer to as a 2-log reduction). Even if we kill 99.9 per cent, we still have 10^3 bacteria. It is a useful effect, but not as dramatic as the original claim sounds.

When we grow a bacterium in a liquid medium (a liquid culture), it goes through several recognizable stages. When the culture is inoculated (that is, a relatively small number of bacteria are put into the broth), nothing much appears to happen for a while. This is the so-called *lag phase*. Essentially, the bacteria are getting used to the change from the resting state in which they have been stored, and they are responding to the availability of food by making all the various components needed for growth. Some genes (those needed for the resting stage) are switched off, while the genes needed for active growth are switched on. We'll look further at what is involved in these switches in Chapters 7 and 8.

When the cell is ready, it will start exponential growth. At the end of the log phase, when it runs out of food, the process is, in effect, reversed. The genes needed for active growth are switched off, and the cell enters *stationary* phase. This is not merely the absence of growth. A number of functions are necessary if the cell is to stay alive in stationary phase, so these genes have to be switched on.

Many bacteria, such as *E. coli*, survive quite well in stationary phase, but not all do. Some will start to die, presumably because they do not have the genes needed to keep the cell alive in the absence of growth. Later on, we will also encounter bacteria that are able to form specialized cells known as *spores*, some of which can survive almost indefinitely without any detectable metabolic activity. These dormant structures are extremely important in a practical sense, as they may be extremely resistant to heat and disinfection. Examples of spore-forming bacteria include the organisms responsible for tetanus and botulism (see Chapter 5).

The conventional way of identifying bacteria in a mixture – such as might be obtained from a clinical specimen such as a wound swab, or from an environmental sample – is to look at its biochemical properties. In other words, what chemicals it can grow on, what products it makes, and so on (just looking at them down a microscope usually doesn't tell us very much, although it can help). This means there is a need to purify individual bacteria from the mixture. This is easier than it sounds, provided the bacteria will grow in the lab.

Instead of putting them in a liquid culture (a 'broth'), we would use plastic dishes which are known as Petri dishes, after Julius Richard Petri (1852–1921), who invented them while working as an assistant to the more famous bacteriologist Robert Koch (1843–1910). Into these dishes, we put a medium which is made solid by adding agar (a jelly-like substance made from seaweeds). The bacteria do not move around on this; they just stay where they land.

If a dilute sample is spread on an agar plate, this will create a random pattern of isolated bacteria. They cannot be seen at this stage but, if the plate is incubated at an appropriate temperature, for 1–2 days for many bacteria, the bacteria will multiply. Since they cannot move, this will produce a small blob of bacteria, known as a colony, at the site where they started. If the bacteria were sufficiently well spread out initially, then each colony will have come from a single bacterial cell. An individual colony can thus be picked off and used to make as many cultures as are required. The result is a pure culture of a specific bacterium from the initial mixture, which may originally have contained a lot of different bacteria. This is actually 'cloning' in the original sense of the word – producing a population of identical individuals, all derived from a single bacterium by asexual reproduction.

Bacteria are often referred to as 'prokaryotes', which means that they do not have a nucleus or other compartments such as mitochondria within the cell (but see the section on Archaea later on). All of the reactions within the cell, including replication of the DNA, expression of the genes, and generation of the energy they need, take place within the cytoplasm of the cell. This should not be taken to mean that the cytoplasm is an amorphous soup – it does have structure, but it is quite subtle.

Other microbes have a cell structure that is much more like those of plants and animals; these are the eukaryotes. They have a nucleus, which contains the chromosomes carrying the genetic material, and mitochondria, which are the powerhouse of the cell in that they are largely responsible for energy generation. Some (especially plant cells) also have chloroplasts, which are responsible for photosynthesis. Mitochondria and chloroplasts are interesting, as they also contain DNA, as well as ribosomes (see below), which are responsible for protein synthesis. Thus, mitochondria and chloroplasts resemble organisms in their own right which have become adapted to an existence within the eukaryotic cell. Indeed, it is believed that this is how they originated (see Chapter 10).

The main groups of eukaryotic microbes include fungi, protozoa, and algae. Including fungi may seem surprising, as fungi are familiar to all of us as mushrooms and toadstools. One fungus, a specimen of *Armillaria* that occupies about 1,000 hectares (10 sq km) in Oregon, USA, is often referred to as the largest known living organism; its weight is estimated at over 600 tons (compared to the 200 tons of a blue whale). Similarly, algae include seaweeds.

Surely these are not 'microbes'? To answer this, it is necessary to look at the cellular structure of these organisms. Many fungi can exist either as single cells or as collections of many cells. Mushrooms normally grow in the soil as a network of filaments, called a mycelium, which is composed of many cells joined end to end. Whether we should think of this as a single organism or as a collection of individuals is debatable. In a true multicellular organism (such as ourselves), there is communication and interaction between individual cells, and also differentiation – each cell (or groups of cells) has a specific function and forms a specific structure. So we have a liver, heart, kidneys, and so on. In the mycelium, there is virtually no differentiation, and only a limited amount of communication. However, when

Figure 1.4 Fungi. (a) The mould *Penicillium multicolor*, growing on an agar plate (content provider(s): CDC/Dr. Lucille K. Georg); (b–d) miscellaneous fungi.

conditions are appropriate, differentiation (and communication) does occur, and the fungus produces fruiting bodies, which are the familiar mushrooms. Some cells produce the different parts of the stalk and the cap, and some produce the spores which enable the mushroom to propagate and spread.

It is worth noting that some bacteria – especially the *Streptomyces*, which are common in soil – also grow as a sort of mycelium and can produce spore-bearing structures, although on a much smaller scale. We will come across *Streptomyces* again, as they are the principal source of naturally occurring antibiotics. In a later chapter, we will also come across some other bacteria that show elements of communication and differentiation, and behave in a way that resembles that of a multicellular organism.

The real reason for including fungi as an example of a microbe is their ability to grow as dispersed single cells. Some, such as the yeasts used for baking and brewing, always grow like that, while many others can be grown as single cells in the laboratory. Similar considerations apply to the algae. However, if we apply this too literally, we have a further problem. Many plants can be grown in the laboratory as cultures of single cells (and subsequently induced to form intact plants). Some animal cells can also be grown in this way (although it is not usually possible to regenerate an intact animal from them). Should we therefore also consider plants and animals as 'microbes'? Conventionally we do not (although for a while the microbiology degree course at my university did include a module on plant and animal cells).

In my list of organisms I included algae and protozoa, but these should really be considered together as a larger group known as protists, because there is a considerable degree of overlap. At one extreme, we have what can be regarded as a typical protozoan, which resembles an animal cell in that it does not have a rigid cell wall, so it is a very flexible organism which moves around by extending its surface in one direction; it typically feeds by simply engulfing part of the liquid around it and digesting whatever it contains. In some cases, protozoa can feed on simple nutrients in the medium, but more interesting is their ability to ingest bacteria. The ability of protozoa to feed on bacteria is an important factor in soil and water ecosystems. These organisms are referred to as amoebae. Many protozoa have more complex structures, including in some cases a 'mouth' and cilia which waft particles into the 'mouth'.

At the other extreme, we have typical algae, which are photosynthetic (they have chloroplasts containing chlorophyll), with a rigid cell wall (and so are similar to

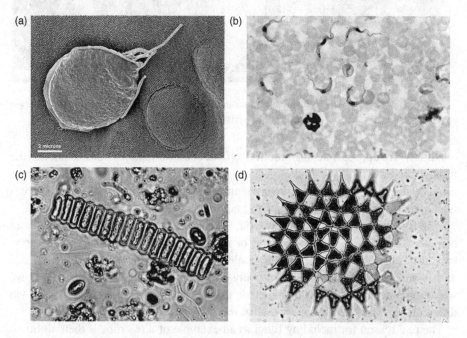

Figure 1.5 Protozoa and algae. (a) Colourized scanning electron micrograph of the protozoon *Giardia muris* adhering to an intestinal epithelial cell (content provider(s): CDC/Dr. Stan Erlandsen); (b) Stained light photomicrograph of the protozoon *Trypanosoma brucei* in a blood smear (content provider(s): CDC/Dr. Mae Melvin); (c) and (d) Green microalgae: (c) *Scenedesmus bijuga* (*Source:* United States Environmental Protection Agency Great Lakes National Program office); (d) *Pediastrum simplex* (*Source:* United States Environmental Protection Agency Great Lakes National Program office).

plant cells). Many of them can move around using flagella. However, the protist group also includes a number of important organisms which do not fit properly into either camp. Especially worth a mention are the diatoms, which have a beautiful glass-like wall made of silica, and the dinoflagellates, which are commonly responsible for the so-called 'algal blooms' that sometimes occur in seawater; some of these produce powerful toxins which have caused severe food poisoning from contaminated shellfish. Other protists have some similarity to fungi. These include the oomycetes, many of which are serious plant pathogens – an oomycete was responsible for the Irish potato famine in the 1840s.

Here it is worth mentioning the organisms that are sometimes, erroneously, referred to as 'blue-green algae'. These are not algae at all but are photosynthetic bacteria, and they are more properly referred to as cyanobacteria. We will come across cyanobacteria in several contexts in this book, as they are very important components of our ecosystem. Indeed, it was the ability of cyanobacteria to carry out photosynthesis, leading to the incorporation of carbon dioxide into living matter and the corresponding release of oxygen, that led to the rise in oxygen levels in our atmosphere which made possible the subsequent evolution of many forms of life as we know it today.

Even more remarkable are the slime moulds. These live in the soil, as single cells, feeding on decaying matter or bacteria, until the food starts to run out. Then, if there are enough of them in a particular niche, they come together as an aggregate and form a tiny slug-like structure, a few millimetres long. Subsequently, the various cells within the 'slug' start to differentiate. Some form a base plate, others contribute to a stalk, and yet others turn into spores within a fruiting body. This cycle between unicellular and multicellular organization makes them a valuable subject for scientists who are interested in the mechanisms underlying development and differentiation.

Clearly, the slime moulds can only do this if there are enough of them present to form all the various bits of the final structure. How do they know when there are enough? This is an example of a widespread phenomenon known as 'quorum sensing'. Each cell secretes a chemical which acts as a signal and is recognized by other cells. If there are enough cells in the immediate neighbourhood, the concentration of this chemical will reach a critical value that permits the subsequent development. We will encounter more examples of quorum sensing later on, as it plays an important role in the way in which some bacteria cause disease, as well as in other phenomena.

The next characters to be introduced are less likely to be familiar. Studies of microbes from various environments, including extreme conditions, such as hot springs and salt lakes, discovered some unusual organisms that, in some ways, resembled bacteria, especially in lacking a nucleus; since they were thought to be primitive organisms, they were dubbed Archaebacteria. However, it was subsequently realized, from comparisons of genome sequences, that they were fundamentally different from bacteria, and that they are actually more closely related to the eukaryotes. So, they are now called Archaea. As well as living in extreme environments, other members of the Archaea occur in diverse situations ranging from waterlogged soils to the gut of animals (including humans) – and some have the significant property of producing methane.

Since the Archaea do not have a nucleus, they also fall within the Prokaryotes which I referred to earlier. Many microbiologists now regard this as misleading, as the Archaea are quite distinct from bacteria, and therefore the term 'prokaryotic' is now out of favour in some quarters.

I want now to turn back to the other end of the scale of sizes. I started with viruses, and the question of whether these are really 'living'. However, viruses do resemble living organisms in having nucleic acid (DNA or RNA) that codes for proteins. The final members of the cast are even more different, in that they have neither DNA nor RNA. These are the prions, which rose to notoriety in spectacular fashion with the advent of BSE ('mad cow disease', or bovine spongiform encephalopathy) and its human variant Creutzfeldt-Jakob disease. The infectious particle consists solely of a misfolded form of a normal protein. The 'replication' of the prion resembles growth of a crystal, in that copies of the misfolded form of the protein come together as aggregates, which act as seeds to induce further misfolding of other copies of the normal protein. The aggregates increase in size and then split to seed more misfolded aggregates. This process is quite different from replication of living organisms, so we should not really consider them as a form of life. However, they can be transmissible, so this, to some extent, justifies an inclusion here.

However, I should emphasize that they are not always transmissible. Transmission of BSE to humans only occurred through eating meat from a cow with BSE, so human to human transmission was not a problem (although a disease called 'kuru' did occur among people in New Guinea with the cultural habit of eating the brains of dead relatives – often, but misleadingly, referred to as 'cannibalism', although it was more of a mark of respect for the dead rather than the image that the word 'cannibalism' conjures up). I will look further at BSE in Chapter 3.

Prions may have an importance beyond their role in BSE and similar diseases. In particular, there are suggestions that the mechanisms behind degenerative diseases such as Parkinson's and Alzheimer's have some similarity to those involving prions – although they do not seem to be infectious.

In subsequent chapters, we will look at various aspects of microbial behaviour – how they cause disease, their activity in the environment, and how we can, and do, use microbes in a variety of ways. But first, I need to cover some basic concepts that will recur throughout the book. Some aspects are also included in Appendix 1, which covers some basic material as well as going rather further into selected topics than is included here.

1.2   Food for microbes

If we think about those bacteria that are able to grow on simple media, the main components needed for growth are sources of carbon and nitrogen, plus an energy source. Many bacteria, such as *E. coli*, can combine a carbon and energy source, using simple organic compounds such as glucose (or other sugars) to serve both purposes. The glucose is broken down through a series of reactions (ultimately to carbon dioxide and water). At several stages in this process, energy is released, in a controlled manner, so that it can be coupled to the synthesis of the wide variety of materials needed for growth – proteins, nucleic acids (DNA and RNA), lipids, and all sorts of other things. For a nitrogen source, *E. coli* is quite happy with an inorganic substrate such as an ammonium salt.

One of the main ways in which the energy is released in a controlled manner occurs by the passage of electrons (see Appendix 1) from one intermediate to another, until it is eventually passed on to a final electron acceptor. The energy released at each step is harvested by coupling the reaction to the production of a chemical known as adenosine triphosphate (ATP). This is the major energy resource within the cell and it is used in many reactions where energy is needed. For many of the bacteria we will be looking at, which are able to grow aerobically (in the presence of air), this passage of electrons along a chain of cytochromes, known as aerobic respiration, involves oxygen as the final electron acceptor, and the chain of intermediates is therefore called the *respiratory chain*. Some bacteria use a similar process even in the absence of air (*anaerobic respiration*), in which case some other substance (such as nitrate) is used as the final acceptor of electrons.

I need to introduce two further chemicals that are central to these, and many other processes. These are the related compounds known as NAD and NADP. Both can act as electron acceptors, which converts them to the reduced forms NADH and NADPH. Conversely, NADH and NADPH act in other reactions as electron donors, which changes them back to the oxidized forms NAD and NADP (oxidation of a substance is equivalent to the removal of electrons from it, and the converse process, reduction, is the addition of electrons – see Appendix 1 for further explanation). So NAD and NADP act as catalysts in many of the oxidation and reduction systems within the cell.

Not all bacteria have a respiratory chain. Instead, they obtain their energy through the process of *fermentation*. Several of the steps involved in the breakdown of a sugar such as glucose can be directly coupled to the production of ATP, without using a respiratory chain. However, this process is much less efficient, that is far fewer molecules of ATP are produced from the fermentative degradation of glucose than can be achieved using a respiratory mechanism. Many bacteria, and other microbes such as yeasts, can also use fermentation pathways under appropriate circumstances, even though they have a respiratory chain.

Other bacteria, such as the photosynthetic cyanobacteria, are even more versatile. They do not need an organic source of carbon because they are able to fix carbon dioxide – that is, they can use carbon dioxide and convert it into organic compounds within the cell. The carbon dioxide comes ultimately from the air, although the form in which they use it is in solution in the water in which they live. This does not provide them with energy; indeed, it requires energy for it to happen. This energy is provided by sunlight, and this is captured, in cyanobacteria, as in green plants, by the pigment chlorophyll. Further information is provided in Appendix 1, but the overall message is that light provides the energy needed for splitting water molecules, liberating oxygen, and for the production of ATP and NADPH, both of which are needed for fixation of carbon dioxide. Surplus ATP and NADPH are used for the production of other material within the cell.

That deals with carbon fixation. The other side of the coin is the release of carbon back into the air. This is done first and foremost by all organisms that carry out aerobic respiration – microbes, animals (including us), even plants and other photosynthetic organisms in the dark. We breathe out carbon dioxide as the waste product from our metabolism. Even microbes that are fermenting their carbon sources rather than carrying out aerobic respiration still emit carbon dioxide – think of the yeasts that are used for making beer. In anaerobic environments, such as the sediment at the bottom of a lake, microbes produce methane, rather than carbon dioxide, as the end product of their breakdown processes (as they also do in the gut of ruminants and, to a lesser extent, of other animals, including us).

The combination of these processes – the fixation of carbon dioxide by photo-synthetic organisms and the breakdown of organic compounds into either carbon dioxide or methane – is referred to as the *carbon cycle*. The balance of this cycle is an important factor in maintaining the composition of the Earth's atmosphere and, of course, it has important consequences for the 'greenhouse effect' and climate change (which I will deal with in later chapters).

There is an analogous cycle operating for nitrogen. The fixation of nitrogen from the air occurs by an energy-requiring process, using the enzyme *nitrogenase*, to produce ammonia (NH_3), which dissolves in water to produce ammonium ions (NH_4^+). This can be oxidized by other bacteria to nitrite (NO_2^-) and nitrate (NO_3^-) – a process known as *nitrification*. These soluble forms of nitrogen can be used by many microbes, and by plants, and are incorporated into amino acids for the formation of proteins and other nitrogen-containing organic compounds, of which the nucleotides that form the nucleic acids are the most important.

The other side of the nitrogen cycle involves the release of this fixed nitrogen back into the air. The decomposition of dead organic matter includes breakdown of proteins to amino acids (using enzymes known as *proteases*). The amino acids are further degraded to release ammonia. We can also add in the nitrogen that is excreted by animals as ammonia, urea or uric acid. The nitrogen cycle is completed by *denitrification*, by yet further microbes in the soil and water, which involves the conversion of ammonia to nitrogen gas through a series of intermediates including nitrate and nitrite. Excess nitrate in the soil, such as may occur through the use of fertilizers, can also be subject to denitrification in this way. I will come back to carbon and nitrogen cycles in Chapter 6.

There are two other elements that are present at significant (although lower) levels in organic matter. These are phosphorus (P) and sulphur (S). Phosphorus is important in nucleic acids, as the link between adjacent nucleotides in the chain, as well as in numerous other compounds such as ATP. It is widely present in the environment as phosphate (PO_4^{3-}) – indeed, it is almost exclusively in this form – which can be used directly by microbes (and other organisms). So we do not need to consider it further in this context, except to say that in the environment, especially in water, phosphate levels may be low, and microbial growth can be stimulated very substantially by addition of phosphate (as occurs in the run-off from agricultural land).

Sulphur is needed in small quantities (it forms just one per cent of the cell's mass), and it is usually taken up as sulphate (SO_4^{2-}). However, reduced forms of sulphur also exist in the environment – hydrogen sulphide (H_2S), elemental sulphur, and in combined forms in rocks and metal ores. Some bacteria are able to use these forms of sulphur by oxidizing them to sulphate. Conversely, other bacteria are able to reduce sulphate to hydrogen sulphide (by anaerobic respiration – i.e. they are using sulphate as the final electron acceptor instead of oxygen). This anaerobic reaction commonly occurs in situations such as the mud at the bottom of a lake, and it is responsible for the foul smell that results if the sediment is stirred up.

Microbes also need a range of trace elements, and they have evolved efficient mechanisms for obtaining these from their environment, so we do not usually need to add them to our culture medium in the laboratory. However, in the environment, especially in the oceans, metals such as iron may be present at levels that are too low for optimum growth of microbes, so growth can be stimulated by addition of minerals (as is the case with phosphate, as mentioned above). Some microbes, especially pathogenic ones, also need some vitamins or other organic compounds to be supplied. In a diagnostic medical microbiology laboratory blood or blood products, or sometimes more specific supplements, are often added to the medium to enable the isolation of these more fastidious bacteria.

It should be realized that this is a rather superficial and generalized account of bacterial nutrition. A full treatment would take a whole book.

1.3   Basic molecular biology

Although this is not a molecular biology book, the techniques and concepts of molecular biology have become so important to many aspects of our understanding of microbial behaviour – not just in biotechnology – that a basic coverage is inescapable. I will try to keep it simple. Some of it is also covered in Appendix 1.

The basic genetic information in the cell is encoded in DNA, in a series of 'bases' or nucleotides, referred to as A, G, C and T. Most bacteria have between 4–10 million bases in their DNA (although some genomes are smaller). Eukaryotes tend to have more – human DNA has a thousand times as much. Eukaryotes also tend to have a number of chromosomes, while bacteria usually have just one DNA molecule (although we will see in Chapter 7 that it is not always so). This DNA is double-stranded, with one strand being a sort of mirror image of the other – for each A in one strand there is a T in the other, and each G is matched by a C. We say that the two strands are complementary. So, if we know the sequence of one strand, we can deduce the sequence of the other. This double-stranded nature is important in copying the DNA. The two strands gradually peel apart, and each strand is used as a template to make a new complementary strand.

The genes in the DNA code for proteins. Each gene is not physically separate; they are just part of a continuous sequence. When a gene is expressed, an enzyme called RNA polymerase recognizes a DNA sequence as a position to start copying one DNA strand into a slightly different nucleic acid called RNA, and it makes an RNA molecule corresponding to a single gene (or, in many cases in bacteria, a group of related genes). It is this *messenger RNA (mRNA)* that is used as the information for production of a specific protein.

Protein synthesis involves structures known as *ribosomes*. These recognize a specific site on the mRNA and start protein synthesis from there. The mRNA is read in groups of three bases (*triplets*, or *codons*), each codon corresponding to a specific amino acid. For example, where there is a CCG triplet, the amino acid arginine will be incorporated in the protein; if the codon is ACG, the amino acid incorporated will be serine. There are 64 possible triplet codons and 20 amino acids, so some of the amino acids are coded for by more than codon. There are also three *stop codons*, which are signals for the synthesis of that protein to stop.

Each gene codes for a specific protein (although that is a simplification). A typical bacterium might have 4,000 such genes. Proteins do a variety of things within the cell. For example, many of them are *enzymes*. These are biological catalysts, and are responsible for all the biochemical systems that occur within the cell, ranging from simple reactions such as an individual step in the breakdown of a sugar molecule to the synthesis of complex structures such as DNA. Other proteins are located in the membrane and are responsible for taking up chemicals from the surrounding medium, such as sugars or other nutrients. They can do this in a highly specific and controlled manner. In some cases, they form a pore through the membrane which will permit the passage of the relevant substance. Since this is a diffusion process, it only works if there is a higher concentration outside the cell than inside. Other proteins can take up substances against a concentration gradient – they pump it into the cell, so this is an energy-requiring process. Conversely, membrane proteins can pump unwanted chemicals out of the cell. Both processes can influence the concentration of an antibiotic within the cell, and can therefore affect the cell's sensitivity to that antibiotic (see Chapter 4).

Another important function that is controlled by specific proteins is the regulation of the cell's metabolic activity. For example, there is no point in *E. coli* making the enzyme β(beta)-galactosidase (which breaks down lactose into glucose and galactose, which is the first step needed for using lactose as a carbon source) if there is no lactose present. Therefore, it makes a protein (a repressor) that binds to a specific

DNA sequence, at the start of the β-galactosidase gene, and prevents it being expressed. If lactose is present, it binds to the repressor protein, and this causes a change in its shape so that it can no longer bind to this DNA site. Thus, the gene is expressed, and the bacterium can start breaking down the lactose. There are a large number of such regulatory proteins, and they play a key role in the adaptability of the cell to different environments.

This is just the basics. Bacterial genetics and molecular biology is much more exciting than this, and we will see in Chapter 7 how our knowledge of molecular biology can explain, and elucidate, the fundamental behaviour of bacteria in important respects, as well as how we can manipulate genes and determine the complete genome sequence of bacteria and other organisms.

2
Microbes and Health

2.1 Microbes in the body

We are surrounded by microbes. They are in the air we breathe, in the food we eat, in the water we drink. And most significantly of all for us, they are within us, in enormous numbers. It has been estimated that the human gut contains perhaps 10^{15} bacteria (i.e. 1 with 15 zeros). Compare that with the number of human cells that makes up the human body, estimated at 10^{13} cells; there are 100 times more bacteria in human intestines that there are human cells in the body. And their diversity is huge – these bacteria come from perhaps 1,000 different species. We then have to add in the bacteria that live in other sites – the mouth, upper respiratory tract, vagina, skin. These bacteria normally do us no harm at all – indeed, they contribute extensively to our normal health, not least in protecting us from infections. How they do this is quite complicated, and not fully understood, but at least in part it is due to competition.

The enormous numbers of bacteria are competing with one another for space and for nutrients, and eventually they reach a more or less stable equilibrium, with each species having found its own niche which it will defend against intruders. In addition, there is a complex and extensive interaction between these bacteria and our own cells, which have the task of allowing the 'good' bacteria to persist but preventing deleterious ones from gaining a foothold – and, especially, trying to keep them all in their right places and not letting them escape into the blood, where even the good ones can be damaging.

The human body is, therefore, an extremely complex ecosystem in its own right. The best example to look at is the gut. Here, the food we eat is exposed not only to a variety of enzymes and other secretions, but also to the microbes (mainly bacteria) that live in the gut. These bacteria play an important role in digesting the food and also, as we shall see, can influence our health in a number of ways.

Understanding Microbes: An Introduction to a Small World, First Edition. Jeremy W. Dale.
© 2013 John Wiley & Sons, Ltd. Published 2013 by John Wiley & Sons, Ltd.

It is important to realize that the bacteria in the gut will also respond to the composition of the food we eat. This is their food as well as ours, and the types of bacteria that inhabit the gut will be those that are best able to use the food we supply them with. If we change our diet, for example by switching from carbohydrate-rich foods to a diet with a high fat content, then the composition of our gut flora will change. Even relatively minor changes in what we eat can alter the competitive balance of microbes in the intestines – so, a short-term tummy upset following a nice dinner with your friends does not necessarily mean they have given you food poisoning.

As a slight digression, it is worth looking at a specific example where the interaction of food and gut microbes causes a problem – namely, lactose intolerance. Some people are unable to drink milk, or eat dairy products, without suffering a variety of intestinal problems such as diarrhoea, wind and stomach cramps. To understand this condition, we have to consider the fate of the natural sugar lactose, which is present in milk. Lactose is a disaccharide – that is, it is made up of two sugars (glucose and galactose) joined together. When babies drink milk, whether it is cows' milk or human milk, the lactose is split into its two constituent sugars by the enzyme β-galactosidase. The glucose and galactose can then be further metabolized and absorbed. However, lactose-intolerant adults do not produce this enzyme, with the result that the lactose remains in the intestine, where it is broken down by gut bacteria, producing copious quantities of gas, which is the main cause of the problem.

An interesting twist to this story is that lactose-intolerance is the normal condition for adults in most parts of the world. The 'normal' situation is that β-galactosidase production gets switched off after infancy. It is only certain groups of people, mainly Europeans, where evolution has altered the mechanism so that β-galactosidase production does not (in most cases) get switched off but continues into adulthood, thus enabling the ubiquitous consumption of milk and dairy products in European culture.

Let's consider the journey that food takes when we eat it. First of all, there is the mouth. The main process here involves enzymes called amylases, which are present in saliva and start to break down starch into sugars (assuming we chew our food properly). There are also many types of bacteria in the mouth – estimated at over 700 species – but their main activities are less than helpful. Some of them form biofilms (plaque) on the surface of the teeth, where they cause caries by fermenting sugars to produce acid. I don't want to discourage anyone from brushing their teeth but, because these biofilms form very rapidly, teeth are again covered with plaque within a short time after brushing. The best defence against caries is to avoid providing the bacteria with a constant supply of sugar.

However, brushing of teeth, if done thoroughly, does clear out the crevices where other bacteria reside. Many of these bacteria in the crevices are extreme anaerobes (they are actually killed by even brief exposure to oxygen), which provides a good example of the extent to which environments can differ over very short distances. In the cracks and crevices, all the oxygen is rapidly consumed by the surface microbes, so those that are further inside the crevice are in an environment that lacks oxygen. It is the activity of these microbes that is responsible for periodontal disease, which is a major cause of teeth loss.

Another unpleasant effect of bacteria in the mouth is bad breath, or halitosis. Some bacteria that are growing anaerobically produce sulphur-containing compounds (e.g. hydrogen sulphide (H_2S) and methylmercaptan), which are responsible for the unpleasant smell.

Carrying on down the gut, the food enters the stomach. Here, it is subjected to proteases (enzymes that break down proteins) and also to hydrochloric acid (HCl), secreted by cells lining the stomach. The acidity of the stomach (with a pH as low as 2) makes it an inhospitable environment for microbial life; indeed, the stomach was considered for a long time to be devoid of resident bacteria. The stomach acidity, therefore, provides an important protection against potential pathogens.

This protective effect is seen most dramatically with the bacterium *Vibrio cholerae,* which causes cholera. Normally, this is killed in the stomach, so we have to take in very large numbers (usually in contaminated water) before we contract the disease. However, if we take some sodium bicarbonate at the same time (which counteracts the stomach acidity), the infective dose is very much lower, You may wonder how we know that. Experiments were carried out in the USA, in 1969, using prisoners as 'volunteers'. There is another unfortunate twist to this story. In many countries where cholera is (still) endemic, many people do not get enough to eat and, if someone is undernourished, their stomach is less acidic. So, as well as being more likely to drink contaminated water, they are more susceptible to this infection.

Clearly, some bacteria can pass through the stomach, otherwise we would never get diseases such as gastroenteritis. How does this happen? Part of the answer is that bacteria can be lodged within food particles that are not fully penetrated by the stomach acidity while the food is in the stomach – it stays there for only 30–60 minutes. Some of the food ingredients also tend to counteract the stomach acidity locally within the particle.

Despite the damaging effects of all that acid, we now know that there are bacteria that do live in the stomach. A bacterium called *Helicobacter* has evolved mechanisms for avoiding and resisting stomach acidity, and it is an important cause of several significant diseases. We will look further at *Helicobacter* and cholera in the next chapter.

From the stomach, the journey of our food takes it into the small intestine (to keep things relatively simple, I will ignore the different sections of the small intestine). This is only 'small' in respect of its diameter (relative to the large intestine) – it is about six metres long altogether, with a highly convoluted lining that has a surface area as big as a tennis court. Although there are bacteria resident in the small intestine (up to 10^6 per ml in the lower reaches), we can consider the small intestine as primarily being a biochemical factory, driven by enzymes which are secreted by the pancreas and by cells in the wall of the intestine. These continue the digestion of proteins and of polysaccharides such as starch, and they also break down other materials in our food, such as nucleic acids and fats – the latter being emulsified by the bile salts from the gall bladder. The other important function of the small intestine is absorption of the breakdown products, which is where its enormous surface area comes into play. Although our food moves swiftly through the small intestine (taking about 2–4 hours), it is here that most of the nutrients pass into the blood stream or the lymphatic system.

It is when our food's journey reaches the large intestine (colon) that things become really interesting from a microbiological viewpoint. We should not ignore other functions of the large intestine, primarily the absorption of water. We pour about seven litres of water per day into the small intestine (including the fluid that we secrete, as well as liquid we ingest). Although much of this fluid is reabsorbed in the small intestine, along with the uptake of nutrients, the large intestine plays a major role in reabsorbing most of the remainder, converting the fluid material emerging from the small intestine into a sluggish mass which moves quite slowly, taking 24–72 hours to traverse the 150 cm or so of the large intestine. Anything that interferes with this uptake of water results in the watery faeces that we know as diarrhoea – most notably with cholera (as we will see later on), where the extensive loss of water is one of the main problems.

The large intestine contains massive numbers of bacteria, estimated at 10^{12} per ml, which means there are a million million bacteria (1 with 12 zeros) in a fraction of a teaspoonful. Traditional bacteriological techniques – trying to grow these bacteria in the lab – only succeeded in identifying a few of the species present. Modern techniques, involving sequencing all the DNA present in faecal specimens (a technique known as metagenomics), have now shown that each person has some 200 different major species of bacteria (and almost certainly many more in smaller numbers that have not been detected). These studies have now been done on hundreds of people and have shown considerable variation from one person to another, so thousands of species have been identified in this way, of which less than 20 per cent had previously been grown in the lab (in a later chapter, we will see that, in environmental samples, the proportion of bacteria that had previously been characterized is even less than this). The vast majority were totally unknown before.

However, the power of this method is such that it is possible, from the DNA sequence, to predict the sequences of the proteins that the bacteria make. By comparison with known proteins from other sources, the biochemical activities of those proteins can in turn be predicted. Thus, it is possible to identify many of the metabolic activities of these novel bacteria, such as the ability to degrade pectin and sorbitol, both of which are present in fruits and vegetables but are poorly absorbed by the body without microbial intervention.

A prominent metabolic activity of gut bacteria is gas production. The bacteria in our gut produce up to one litre of hydrogen per day. Other bacteria use this hydrogen and convert it either to methane (in about 40 per cent of people) or to hydrogen sulphide (in the other 60 per cent). The reason for the difference lies in the composition of the gut flora and, in particular, two groups of bacteria: the methanogens (which produce methane) and the sulphate-reducing bacteria (SRBs), which produce hydrogen sulphide. These seem to be, to an extent, mutually exclusive. Some people have a lot of methanogens and few SRBs, so they produce methane. In others, the converse is true, and the consequence is hydrogen sulphide.

Bacterial metabolism is not uniform in the large intestine. When the food first enters the large intestine, the bacteria will start by attacking the more easily digested carbohydrates, producing useful nutrients which we can absorb. Although absorption in the colon is less efficient than that in the small intestine, enough happens to be significant.

Further down the colon, protein breakdown becomes more important. This is relevant because many of the products of protein breakdown – such as phenols, ammonia, and sulphur-containing compounds – can be toxic or carcinogenic if present in excess. Our gut flora is capable of producing at least ten different carcinogens. Other products have a variety of pharmacological effects on our body. A prolonged transit time in the large intestine, such as occurs during constipation, or with a low carbohydrate/low fibre diet, can have a variety of physiological effects, ranging from increased body temperature to depression, schizophrenia or colorectal cancer. Thus, the traditional insistence on 'regular bowel movements' is not just an old wives' tale, but does have a potential significance for health. This factor may also account, at least in part, for the link between diet and colorectal cancer, and the apparent beneficial effect of a high fibre diet (to avoid alarming anyone unnecessarily, I should say that there are many other factors involved, so constipation does not mean someone is bound to get cancer or any of the other effects mentioned).

There is an increasing body of evidence that microbial activity in the gut can contribute in similar ways to a variety of diseases. For example, one study has shown that patients who have had a heart attack have high levels of various chemicals in their blood, including choline and trimethylamine-N-oxide (TMAO). These originate from lecithin (phosphatidylcholine), which is present to varying extents in much of our food (ranging from egg yolks to soy beans) and is a major constituent of cell membranes (the dietary supplement sold as 'lecithin' may contain a lot of other things as well as phosphatidylcholine). Lecithin is split to release choline, which is then metabolized by gut bacteria to trimethylamine (a gas that smells like rotten fish, and is actually responsible for that smell), which is, in turn, converted to TMAO by the liver – and there is other evidence that TMAO is associated with heart disease.

The role of the gut bacteria is confirmed by experimental studies in mice. When fed high levels of choline, they develop atherosclerosis (furring of the arteries), but this does not happen with germ-free mice. A curious twist is that choline is actually an essential nutrient (deficiencies can cause fatty liver disease) and is sold as a dietary supplement. Decisions as to what is good or bad for us are complicated!

Further evidence of a potential connection with disease comes from metagenomic studies (as described above) of people with Crohn's disease or ulcerative colitis – two conditions where the causes remain unknown. In both cases, there were differences observed in the microbial content of the subjects' faeces, compared with healthy controls.

However, it is important to realize that 'correlation does not imply causation' – in other words, the fact that the microbial content of faeces is different in people with these conditions does not necessarily mean that the different microbes *cause* the condition. The association between these diseases and differences in the gut flora could equally mean that the environment in the gut may have changed so that different bacteria can flourish. Or both factors could be a response to some other, as yet unknown, difference – diet, for example. The same caveat has to be applied to other studies that have shown differences in the bacterial content of the gut between lean and obese individuals. It is tempting to suggest that some people have bacteria in their gut that are more efficient in liberating the nutritional properties of the food that we eat, and that altering the gut flora might provide a way of countering this – but this is still speculative. Some of these examples are considered further in Chapter 10.

We are in danger here of being too negative about our gut flora. The positive side is very important. With experimental animals, such as rats, it is possible to breed and raise them in such a way that they have no bacteria in their gut (or anywhere else). These 'germ-free' (or gnotobiotic) animals grow more slowly than conventionally reared rats. When they are subsequently exposed to a 'normal' environment, so that they encounter bacteria again and their guts become colonized, they start to put on weight more rapidly. Also, we know that giving antibiotics over an extended period, so that the bacterial content of the gut is diminished, makes us susceptible to infections such as *Clostridium difficile*, which are able to exploit the gap left by the eliminated bacteria.

The importance of bacteria in the gut is exploited by the manufacturers of so-called 'probiotic' foods, such as yoghurt containing 'friendly bacteria'. However, if the gut is such a competitive environment, what will be the fate of the ingested bacteria? If we already have a healthy normal flora in our intestines, the chances are that the added bacteria will not establish themselves – we already have a lot of friendly bacteria in our gut, and the enormous numbers of bacteria already there make it a highly competitive environment in which intruders are not welcomed. The probiotics may have some benefit in helping to re-establish a balanced environment in our intestines if we are suffering from an upset tummy, but they are arguably of doubtful benefit if we are already well.

So far in this chapter I have just been considering the normal flora of humans. There is one feature of the normal flora of other animals that needs special consideration, which is the ability to digest cellulose. Cellulose forms up to 70 per cent of plant material and it is remarkably resistant to digestion. Only microbes can break down cellulose. You may then wonder how animals such as cattle and sheep can survive by eating grass – we can't, but they can. The secret lies in the activity of microbes within special chambers that form part of their digestive system.

Ruminants, such as cows and sheep, have an additional chamber, the rumen, into which ingested material goes before entering the main gastro-intestinal tract. Microbes within the rumen convert the cellulose into other compounds that can be absorbed by the intestines and used by the animal. This requires the plant material to be thoroughly broken down, mechanically, and well mixed, which is achieved by regurgitating it and chewing it repeatedly – the familiar action of 'chewing the cud'. There is a rich diversity of organisms involved, and consequently differences in the products formed. The main useful products are short fatty acids such as acetate.

Microbial activity also helps to supplement the relatively low levels of essential amino acids in the diet of herbivores (amino acids are the constituents of proteins; animals can make some of them, but the others – the essential amino acids – are required in their diet). Less useful products are hydrogen and methane. There are some 1.5 billion cattle on the planet, generating over 100 million tons of methane per year. Since methane is a potent greenhouse gas, it is often said that cows play a major role in the greenhouse effect and, indeed, cattle are estimated to produce about 20 per cent of global methane emissions. However, this represents less than 5 per cent of total greenhouse gases (as CO_2 equivalents), so the effect is actually quite small.

Not all grass-eating animals are ruminants. Rabbits, for example, do not have a rumen, but they do have a specially adapted fermentation chamber in which microbes are able to digest cellulose. However, in this case, it is lower down in the digestive system, beyond the small intestine. This poses something of a problem. The large intestine is relatively poor at absorbing nutrients, so much of the potential food is wasted. Some of these animals, including rabbits, circumvent this difficulty by eating their own faeces. We may regard this as a disgusting habit, but it is quite natural and important for rabbits, as it enables the microbial products made in the lower gut to be absorbed by the small intestine. The faecal pellets are actually a valuable food source for rabbits.

There's another twist to this story, involving a very different animal, and a very different process. Some species of ants obtain their food by cutting off bits of leaves (hence, they are known as 'leaf-cutting ants'). As with other animals, they are unable to digest the cellulose by themselves, so they employ microbes to do it for them. They do this by cultivating fungal 'gardens' within their nests. They feed these gardens with the bits of leaves that they bring back; the fungi digest the cellulose for them, and the ants get their nutrients from the fungi. This is a very specific mutualistic association; the queen carries a pellet of the specific fungus in a specially adapted pocket in her mouth and, after mating, she uses this to start off a fresh garden in her new nest. The fungus and the ant are mutually interdependent.

Intriguing as this association is, it is far from the end of the story. The mutualistic fungi are, themselves, host to another fungus which is parasitic and can kill the host fungus by secreting chemicals which break down the mycelium, liberating nutrients that are used by the parasite. To counter this problem, the ants carry, within special modifications of their exoskeleton, filamentous bacteria of the actinomycete group, which produce antibiotics that inhibit the growth of the parasite. So there is a four-way association – the ant, the mutualistic fungus, the parasitic fungus and the antibiotic-producing bacteria.

There are over 200 species of ants in this group. Not all of them are leaf-cutters – some use plant detritus or other material – but they all cultivate fungal gardens, all the fungi are susceptible to parasitic fungi and all the ants carry antibiotic-producing bacteria. The species differ in each case, but the general story is the same. This indicates that this four-way association is extremely ancient, and it has been estimated that it originated over 50 million years ago. They have been evolving together ever since.

2.2   Defences against infection

As I said at the start of this chapter, we are constantly exposed to large numbers of all sorts of microbes. How do we protect ourselves against them? We have several lines of defence, of which immunity (antibodies and so on) is only the last one. This is a very complex subject, but, to simplify it, I will divide these protective mechanisms into three classes: external barriers (mechanical and chemical), innate immunity and adaptive immunity.

The external surface of the body (bearing in mind that, anatomically, the space within areas such as the gut and respiratory tract is actually outside the body) is bounded by one of two types of structure: the skin or the mucous membranes.

Skin is a very inhospitable environment for microbes. It is, in general, too dry for most microbes, and the salty nature of sweat sees off most invaders. Furthermore, antimicrobial peptides known as defensins are secreted by skin cells, as well as by epithelial cells in other sites (many other organisms, including plants, invertebrates and fungi, produce similar compounds.) In addition to these defences, any microbes that do attach to the surface are likely to find themselves shed into the surrounding dust, as the cells in the surface layers are constantly being discarded and replaced from underneath. So, despite continual exposure to microbes, there are comparatively few microbes that can colonize the skin. *Staphylococcus* is the most notable of these – all of us carry them, but only a minority of people carry the potentially pathogenic *Staph. aureus*.

Moister areas, such as the groin and between the toes, have a richer flora. One consequence of this richer flora is the smell that can arise. Feet can smell like cheese precisely because the microbes there produce the same chemicals as those that occur by microbial action in cheese, including the sulphur-containing methanethiol. However, very few of these organisms can penetrate the skin barrier and set up an infection. This situation changes if there are breaks in the skin, such as wounds or insect bites, which provide a route for setting up a wide range of diseases.

Although few organisms can colonize the skin, that doesn't stop other microbes from being carried temporarily. Washing of hands, even with bactericidal soap, does not eliminate the resident microbes, but it does help to remove the transient

Figure 2.1 The mosquito *Aedes aegypti* feeding on a human host. This mosquito is the main carrier for dengue fever. (CDC content provider(s): CDC/Prof. Frank Hadley Collins, Dir., Cntr. for Global Health and Infectious Diseases, Univ. of Notre Dame Photo Credit: James Gathany)

contaminants, and it reduces the numbers of resident organisms on the surface (rather than in deeper sites), so in turn reducing the risk of infecting other people.

The mucous membranes are a different matter altogether. Their roles demand that they are moist and flexible, so they are more subject to damage as well as to colonization. Apart from the protective role of the normal flora of many of these sites, we have evolved a variety of ways to protect ourselves. Let's look at just one example – the respiratory tract.

We constantly breathe in, and out, large numbers of bacteria. What happens to them? Here, a most elegant defence comes into play. The airways are lined with two sorts of cells. One sort produce a sticky substance known as mucus, which traps anything that lands on it. The second type of cell has tiny hair-like projections called cilia, which beat in concert in a constant pattern, a bit like a Mexican wave, wafting the mucus up to the back of the throat. There is a similar system in the nose. In both cases, this results in any trapped particles, including inhaled bacteria, being expelled.

The consequences of anything going wrong with this system are all too familiar. Viral infections can cause the cilia to stop beating, so the mucus accumulates in the nose, producing the familiar stuffed-up feeling. The same happens with dry environments, or as a consequence of air pollution (including smoking). Alternatively, viral infections can result in excessive production of mucus, so causing a runny nose.

This defence mechanism is normally very effective. How, then, do bacteria cause infections such as pneumonia (the commonest cause of death due to bacterial infection in the UK and similar countries)? This would require the bacteria to get right down into the lungs, so having to escape this defence system.

There are two answers to this. The most obvious one is that something has gone wrong with our defences, so any viral infection able to disrupt this defence mechanism, such as a cold or influenza, can lead to a more serious bacterial infection. Smoking acts in a similar way – heavy smokers are more prone to respiratory tract infection. We refer to these as predisposing factors, as they make us more susceptible to infections to which we would normally be resistant. A further predisposing factor is cystic fibrosis. This condition is due to a defect in a protein known as the cystic fibrosis transmembrane conductance regulator (CFTR). This protein regulates the transport of chloride ions across the membrane and conse-quently affects the transport of water out of the mucus-secreting cells (since the transport of chloride ions requires the simultaneous movement of water). A result of this defect is that the mucus is too thick and heavy for the ciliated cells to remove it properly. Because of this, people with cystic fibrosis are especially susceptible to respiratory tract infections.

Another factor is more complicated. Most of these respiratory tract infections are caught from someone who is already suffering from this disease, so we need to start with the source of infection. When we cough or sneeze, or even when we just talk or breathe, we expel a fine mist of suspended particles, some of which contain the bacteria (or viruses) that can cause infection (assuming we are infected). The largest of these droplets will fall to the ground (or other surfaces) very quickly, and are of less concern – although they may persist in the dust and cause problems subse-quently. In any case, we are less concerned about these large particles as they will be trapped very efficiently by our mucus if we inhale them.

But what about the smaller droplets? These have a very large surface area compared to their size, so any moisture they contain will evaporate very quickly. So, they start off small, and rapidly get smaller still, producing what are known as 'droplet nuclei'. These are only a micrometre or two in diameter, and they will remain suspended in the air almost indefinitely. Furthermore, such small particles are not trapped efficiently by the muco-ciliary system so, when we breathe in such particles, they can get right down into the lungs, beyond the reach of the muco-ciliary defences.

All is not lost, however. Waiting for them, in the furthest reaches of the airways (the spaces known as the alveoli), is a cell (macrophage) whose job it is to engulf any particles that get that far. This is usually very efficient, with two exceptions. Firstly, some people – including the elderly – do not have very effective immune systems, which reduces the ability of the macrophages to dispatch the invading bacteria. Significantly, it is among the elderly that bacterial pneumonia is a major killer.

Secondly, some bacteria have evolved ways of resisting the killing action of the macrophage. Notable among these is the bacterium *Mycobacterium tuberculosis*, which causes TB. This actually *likes* being inside macrophages, which it finds a nice, comfortable environment, free from other aspects of the immune system, and the bacteria will survive and grow there. The organism that causes Legionnaire's disease (*Legionella pneumophila*) is another example although, in this case, it also needs an impairment of the muco-ciliary system so that larger numbers can penetrate to the alveoli.

Even organisms that are usually killed by the macrophages can set up pneumonia if they reach the alveoli in great numbers. As mentioned previously, viral infections, such as influenza, interfere with the muco-ciliary defence mechanism, which allows bacteria to penetrate to the alveoli. This accounts for the occurrence of secondary pneumonia as a consequence of such viral infections.

One question of practical importance is raised by this discussion. Does wearing a face mask protect against infection? It *is* possible to filter out small particles such as droplet nuclei, containing bacteria or viruses, but this requires a filter with very small pores. Such a filter causes a lot of resistance to the passage of air, so high pressures are needed to force air through it. Filters like this are used in safety cabinets where work with dangerous pathogens is undertaken, in order to make sure that the air pumped out into the environment is entirely free of pathogens. In such a situation, it is possible to use powerful pumps to force the air through the filter. But we cannot breathe through such a filter – the pressures needed are too great, and in any case the mask doesn't fit tightly enough to the face, so air would leak around the sides. So why do surgeons and other medical staff wear surgical masks? The real purpose is to protect the patient. As described above, the particles we expel when we breathe or talk are much larger (the water hasn't yet evaporated), so they are more readily trapped by a mask with a larger pore size.

We can also look at the question of why respiratory tract infections are more common in winter. This has nothing to do with getting cold and wet (and so 'catching a cold'). The real point is that in winter we tend to huddle indoors, with the windows closed, so any pathogens in the air have much more opportunity to infect others. Thus, respiratory tract infections are more common in winter. This contrasts with some other types of infection, notably food poisoning, where most forms are more common in summer. These infections are due to bacteria multiplying in the food, which happens more readily in warm weather (see Chapter 5).

If a microbe manages to evade the first lines of defence, and gets into the blood stream, it is then attacked by a variety of mechanisms known collectively as *innate immunity*, to distinguish it from the more familiar *adaptive immunity*. The latter requires the body to recognize a specific microbe and to produce a specific response to it. As this needs the cells involved to multiply, it takes time to develop. We need a quicker response to kick in while that is happening.

One arm of the innate immune system involves cells in the blood stream known as neutrophils. These are phagocytic, i.e. they engulf and digest anything that shouldn't be there. This is an extremely fast process that will mop up any bacteria that get into the blood in a matter of minutes. So, for example, if we brush our teeth reasonably vigorously, or bite into a crisp apple, the mucous membranes in our mouth and gums will suffer minute amounts of damage, allowing bacteria from our mouth to enter the blood. This is not a problem, as they will be eliminated rapidly.

It has been more recently recognized that a degree of specificity is conferred on this system by the ability of phagocytic cells to recognize certain chemical signatures, known as pathogen-associated molecular patterns (PAMPs) that occur only in microbes (bacteria or viruses), such as structures that form part of the bacterial cell wall. The PAMPs are recognized by special structures called Toll-like receptors (TLRs). When the TLR binds to a specific PAMP, it sets off the expression of genes that are responsible for activating a defensive mechanism that stimulates the cells of the immune system, resulting in inflammation. A series of different TLRs exist, which recognize different signatures, each of which is present in a wide range of bacteria (or, in some cases, viruses). Therefore, it does not matter what microbe it is – the innate immune system is ready to deal with it.

A further defence system comprises a set of proteins known as the complement system. This system is activated by substances on the surface of bacteria, and it acts in a cascade – that is, the first protein activated causes the second protein to become activated, and so on. This amplifies the effect very considerably. The main relevant effect of the complement system is to kill the invading bacteria by lysis, which means the bacteria are split open. Complement also interacts with the adaptive immune response, and is involved in inflammation. Still further protection is provided by antimicrobial peptides known as defensins (as referred to earlier), which act by making holes in the walls of many bacteria.

Yet another defence arises from the fact that the iron in the blood is firmly locked up by being bound to a protein known as transferrin. Our own cells know how to release the iron that they need but, unless bacteria can do the same, they will be prevented from multiplying due to a lack of iron. Bacteria, like all cells, need iron; it is an essential component of a number of key enzymes in the cell.

This is still far from the complete story, but these examples demonstrate the power of the innate immune system to deal with invading microbes – except those that are able to avoid or resist these defences. The ability to evade these defences is one of the characteristics that makes certain microbes pathogenic.

If the invading microbe manages to evade all the defences described so far, the final, and very powerful, line of defence is the adaptive immune system – which is what we normally mean when we talk about 'immunity'.

At the simplest level, when a foreign protein – an *antigen* – gets into our blood stream, it will be recognized by another protein – an *antibody* – that is made by specialized cells, called B cells, in our body. The antibody combines with the antigen, neutralizing any effects it might have, and the combination is then removed and destroyed. There are a vast number of different B cells, each producing different antibodies, and furthermore they are capable of additional variations to multiply the potential repertoire, so enabling us to recognize any foreign protein we might encounter (we can produce antibodies to all sorts of antigens, not just proteins, but, just for the moment, let's limit it to protein antigens). We also make antibodies at mucosal surfaces – in the mouth, the respiratory tract. and in the gut, not just in the blood.

Once an antigen has been recognized, this stimulates the multiplication of the B cell that made it so that we can produce larger amounts of that antibody. Our bodies remember this encounter, so that next time we meet the same antigen, we can produce large amounts of the necessary antibodies much more quickly. We don't go on making it all the time – once the antigen has gone, the antibody levels subside – but we remember and will respond quickly next time round. Vaccination works by imitating infection, but without the symptoms. Thus, when we are really infected with that pathogen, our immune system remembers the encounter with the vaccine and is consequently ready to deal promptly with the infection.

The production of antibodies is often sufficient for immunity to viruses, as antibodies alone can often neutralize a virus. They can also protect us very effectively against those bacterial diseases that are due to the formation of protein toxins, which are also inactivated effectively by combination with an antibody. This includes diphtheria and tetanus, where we use vaccines to induce antibodies against the toxins (see Chapter 4). In both cases, the symptoms of the disease are due to a toxin that is secreted by the bacteria, and antibodies can bind to and neutralize those toxins. For most bacterial diseases, however, and (even more so) for diseases caused by eukaryotic organisms such as protozoa and multicellular parasites, antibodies do not provide much, if any, protection, and in some cases can be quite damaging. For protection against bacterial infections, we often need a different sort of immunity known as cellular immunity.

Cellular immune responses are much more complex, and they involve the inter-action of different sorts of cells, mediated by chemical messengers known as cytokines. We have already met macrophages, which are an example of phagocytic cells that are able, sometimes, to destroy bacteria that they engulf. The killing action of macrophages can be potentiated by cytokines produced by a different class of cells known as T cells. T cells are, in some ways, similar to B cells in that they can recognize, and be stimulated by, specific antigens. The T cells can therefore also remember meeting these antigens and can respond more effectively next time round.

There exists a variety of T cells, which have different effects, including some T cells which kill other cells directly. Others produce different cytokines, some of which stimulate B cells to produce more antibody, which may not be the required response. The production of the cytokines may actually be quite damaging, producing fever and other symptoms (the extensive weight loss that often accom-panies tuberculosis is due to a cytokine called tumour necrosis factor, for example). So, stimulating the T cells may not result in a protective effect; the effect may be useless or even counterproductive (I should point out that this is a highly simplified version of a very complex phenomenon).

It is, at present, very difficult – if not impossible – to predict what effect a new bacterial vaccine would have on the cellular immune response. This is one reason why there are so few widely used vaccines against bacterial infections. The other reason is that we have a battery of antibacterial antibiotics available for the treatment of bacterial infections, so there is not the same overpowering need for vaccines.

In a later chapter, I will deal with the ways in which we can use vaccines to protect us against infectious diseases, as well how antibiotics work. First, I want to consider some examples of the diseases themselves.

3
Microbial Infections

3.1 Diseases of the past

If we turn the clock back 150 years or so, to the mid-19th century, many serious infectious diseases were an everyday matter of life – and death – in the UK and similar countries. Three examples will show the effect – cholera, tuberculosis and (going rather further back) plague. Note that although I have headed this section 'Diseases of the past', this only applies to countries like the UK. On a worldwide scale, these diseases are still very much with us.

During the 19th century, Britain was affected by several epidemics of cholera. Following the Industrial Revolution, there had been a massive movement of people into the cities. For example, London expanded from some 300,000 in 1700, to 800,000 in 1801 and to 3.3 million by 1871. This posed an enormous problem for the provision of water, and the disposal of sewage. Many of the population relied on relatively shallow wells for their water. Where piped water was available, it was often drawn from rivers, which were heavily contaminated by sewage in the absence of the main drainage systems that we regard as commonplace today.

This sounds like a recipe for disaster – as, indeed, it was – but we have to remember that the nature and origins of diseases such as cholera were still unknown at the time. The 'germ theory' of disease (principally ascribed to the work of Louis Pasteur), which held that disease was caused by tiny organisms invisible to the naked eye, was only just being developed in the mid-19th century and was still far from generally accepted. Many preferred to believe in the 'miasma' concept, which held that these diseases originated from the bad air that was common in the overcrowded conditions in the cities. To its adherents, this concept seemed obvious. Diseases were rife in the slums, where one only had to sniff the air to smell the disease potential. We now know that, for diseases like tuberculosis which are spread through the air, there was some truth in this, although it was not the smell that was

---

Understanding Microbes: An Introduction to a Small World, First Edition. Jeremy W. Dale.
© 2013 John Wiley & Sons, Ltd. Published 2013 by John Wiley & Sons, Ltd.

the cause. However, it was not true for cholera, for which John Snow established that it was spread by water supplies.

Snow was born in 1813 and was apprenticed to a surgeon at the age of 14. After some years as an apprentice and assistant in the north of England, he enrolled as a medical student in London and became a member of the Royal College of Surgeons in 1838. It is his interest in the nature and spread of infectious diseases, especially cholera, that has earned him a place in history. During the epidemic of 1854, he observed that there were large numbers of cases of cholera in a part of Soho, in central London, clustered around a pump in Broad Street (now known as Broadwick Street). In the absence of any action by the authorities, he took matters into his own hands and removed the handle of the pump, so forcing the inhabitants to get their water from other, less contaminated, sources. The number of cases then dropped. Well, that is the story, but in fact the outbreak had probably run its course by that time, and the number of cases would have dropped anyway.

A more significant piece of work by Snow, which he referred to as his 'Grand Experiment', was a study of the number of deaths from cholera in a part of South London. This area was served by two water companies, the Lambeth Waterworks Company and the Southwark and Vauxhall Water Company. Their pipes ran side by side down each street and some people got their water from one and some from the other. He painstakingly mapped out which water company served each house, and related that to the number of deaths from cholera. This showed that there was a much higher risk of death from cholera if people got their water from the Southwark and Vauxhall Water Company.

This type of analysis, comparing the rates of infection (or death) in two groups of people who are as far as possible identical in all things apart from the one risk factor we are interested in, is a core procedure in modern epidemiology. The study earned Snow the title of the 'Father of Epidemiology'.

The actual figures give us pause for thought. The study covered seven weeks in July and August of 1854. During that short period, in this part of London alone, there were 499 deaths from cholera, 419 of them amongst users of the Southwark and Vauxhall water. Taking into account the numbers of users, the death rate per 1,000 was 4.2 for the Southwark and Vauxhall water and 0.5 among users of the Lambeth company's water.

Important as Snow's Grand Experiment was scientifically, he did not single-handedly remove cholera as a problem in London. Indeed, by the time of his study, measures were already in place that would drastically reduce the problem.

In 1834, Edwin Chadwick, who was employed by a Royal Commission into the working of the Poor Laws, wrote a report recommending reform of the system. He was appointed secretary to the new Poor Law Commissioners, but he fell out with his employers over the way the system was administered. While still employed in this role, he wrote and published (at his own expense) a treatise (1842) with the evocative title, *On the Sanitary Conditions of the Labouring Population.*

In 1848, he was appointed Sanitation Commissioner under the new Public Health Act. His period in that role was not entirely a happy one, as he continued to ruffle too many feathers to achieve much of what he aimed for, but there was one highly significant advance made. In 1850, he recommended that London's water should be obtained from cleaner sources, and this was implemented in the Metropolitan Water Act of 1852. The Lambeth Waterworks Company must have seen this coming because, in that year, they moved their intake from Hungerford to the cleaner waters of Thames Ditton. This explains why there were fewer cholera deaths amongst people getting their water from that company; the Southwark and Vauxhall Company was still drawing its water from the Thames at Battersea. They moved their intake upstream to Hampton in 1855.

As well as the controversy regarding the source of infection, we also have to remember that, at that time, no organisms had yet been identified as causing infections. Microbes had been seen as early as the 17th century (by Antonie van Leeuwenhoek – see Chapter 1), but it was not until the late 19th century that the first real proof was obtained that they could cause disease. They key person we associate with this was Robert Koch, who established a set of principles that are still referred to as the ideal way of establishing such a link. Koch's postulates can be summarized as requiring it to be shown:

1. that the organism could be isolated from a patient;

2. that it could be grown in pure culture;

3. that the culture would produce the same condition in an experimental animal; and

4. that the same organism could be recovered from the infected animal.

It should be noted that for many well established causes of diseases, it is not possible to fulfil all of these postulates. For example, the bacterium that causes leprosy cannot be grown in the laboratory, so failing at the first hurdle – but that does not prevent scientists from arguing about the need to demonstrate them in controversial diseases. Using this approach, Koch identified the bacterium (*Vibrio cholerae*) that causes cholera, in 1883.

This highlights one of the strengths of an epidemiological analysis, such as that carried out by John Snow. It is possible to track down the source of a disease, and hence take decisions about how to prevent it, without knowing what causes it – and John Snow did not know about the bacterium that causes cholera.

During the second half of the 19th century, there was a parallel development that would also contribute to getting rid of diseases that spread in this way, and again it was not based on knowledge of how these disease were spread. The amount of sewage and other rubbish that was deposited in the Thames had made it, in effect, an open sewer. As a consequence, it became extremely smelly and, on at least one occasion (known as the 'Great Stink', in 1858), sittings of Parliament had to be suspended because the smell was so bad.

An engineer called Joseph Bazalgette planned and pushed through a massive programme of public works – including 134 km of main intercepting sewers and 1,800 km of street sewers – that created, for the first time, a centralized sewer system for London. The main element of this system involved sewers, built behind embankments alongside the Thames, to intercept the flow of drains and under-ground rivers – so, as we walk along these embankments now, we should remember their original purpose. Much of Bazalgette's system is still in use today. However, modern treatment of the sewage only came later; Bazalgette's system still involved dumping it into the Thames – untreated, but much further downstream.

Before finishing with cholera, I should say something about the disease itself. The bacterium that causes it, *V. cholerae*, produces a toxin in the large intestine. This toxin interferes with the uptake and secretion of chloride ions, and consequently with the uptake and secretion of water (since this accompanies the chloride ions). The result is a massive outpouring of watery diarrhoea (up to 20 litres per day). The resulting dehydration (and the accompanying electrolyte imbalance) is what kills. It is futile giving the patient water to drink – it cannot be absorbed.

However, knowledge of the underlying mechanism does provide us with a simple solution. There are alternative pathways for water uptake in the large intestine, especially accompanying absorption of carbohydrates. These are not affected by the cholera toxin. So, drinking a solution of sugar will provide effective rehydration if started early enough. It does not remove the infection, but it keeps the patient alive while the bacterium is washed out of the system. Oral rehydration tablets are a valuable part of our luggage if travelling in a cholera-endemic area – and they are also useful for dealing with other causes of gastroenteritis.

Cholera is now unknown in Britain, but this not due to the advent of vaccines and antibiotics is, but to the public health and engineering work carried out by the Victorians, based entirely on principles other than knowledge of the organisms and how they spread. In many parts of the world, much of the population does not have access to safe drinking water. This not only creates a risk of outbreaks of cholera, which is endemic (i.e. continually present to some extent) in countries such as Bangladesh, but also produces a constant threat of many other diseases. Diarrhoeal disease, due to a variety of organisms – bacteria, viruses, protozoa – is one of the major causes of death in many countries, especially in children; some 1.5 million children die of diarrhoeal disease each year. I will come back to gastroenteritis and other food-borne diseases in Chapter 5.

Tuberculosis is our second example. This has been around for a very long time. The bacterium that causes it (*Mycobacterium tuberculosis*) has been detected in ancient bones dating back 9,000 years. But to look at the real impact of TB as a major killer, we again have to pick up the story in the 19th century. At the start of that century (and probably dating back to at least the start of the 18th century), TB was rife in the UK. It is estimated that 25–30 per cent of deaths in the population were due to TB. The likely reason for the upsurge in TB was the large-scale migration to the cities in the Industrial Revolution. The dramatic increase in the size of the cities at that time (as mentioned previously) resulted in a lot of poor quality and overcrowded housing. TB, as a respiratory disease, spreads readily in over-crowded conditions.

The literature of that period is full of examples, often of tragic heroines dying romantically (think of little Nell in *The Old Curiosity Shop,* Smike in *Nicholas Nickleby* and the central characters in operas such as *La Traviata* and *La Boheme* – although death from TB was far from that romantic ideal). The prevalence of TB in literature is in part due to the dramatic opportunities presented by a disease that progresses slowly, with the subject gradually wasting away. Other infections, such as cholera, although represented in literature (ranging from *The Secret Garden* to *Death in Venice*), cause death much more quickly and in a much more unsavoury fashion.

It also reflects everyday experience. Many of the writers, poets and artists themselves suffered from TB, notable examples being John Keats, Robert Burns and Frederic Chopin, with the history continuing into the 20th century (e.g. DH Lawrence, George Orwell). The list could be very much longer. One of the most famous examples is the Bronte family (Figure 3.1). The first two daughters died in childhood; Emily (*Wuthering Heights*) and Anne (*The Tenant of Wildfell Hall*) also died young. Charlotte survived longer, producing several great novels (notably *Jane Eyre*), but eventually she also succumbed. This is far from a unique family history.

Since then, TB in the UK has declined markedly, so that today it is relatively rare. We hear a lot about the current 'resurgence' of TB, but it is important to put that in context. There has been an increase in the number of cases, but this is quite small compared to the overall long-term decline. In 2009, there were 9,040 cases in the UK, which was the highest level for 30 years. This is not good news – but in 1913, there were nearly 120,000 notified cases in England and Wales.

The reasons for that decline are less obvious than that of cholera. First, we need to think about vaccination and antibiotics. There is a vaccine, BCG (see Chapter 9), which was introduced in 1921, although not widely used in the UK until the 1950s. BCG is an extremely safe vaccine and is given to millions of people worldwide (mostly children, soon after birth) every year, but there is still much doubt over its effectiveness. A trial in the UK in the 1950s showed it to be 80 per cent effective in preventing TB, but several other trials, especially two large-scale ones in South India, appeared to show it had no effect at all in preventing adult respiratory TB. It is still used in many countries where there is a lot of TB, partly because it does seem to be effective in preventing TB in children, and especially a particularly nasty form of the disease known as tuberculosis meningitis. There is also evidence that it has contributed to the decline in the incidence of leprosy, which is caused by a related bacterium, *Mycobacterium leprae.*

Whether it works or not, we do know that it is not the reason for the decline of TB in the UK, since the disease was decreasing in incidence long before BCG was

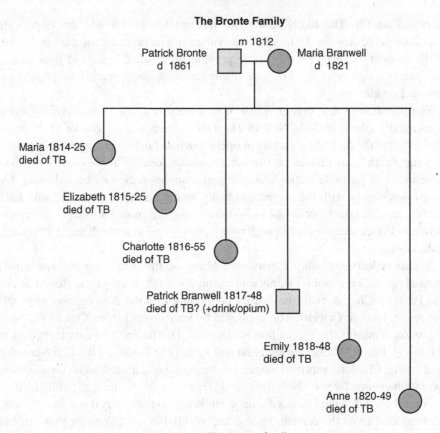

Figure 3.1 The Bronte family.

available. We also know that the decline was not due to antibiotics, as the first effective treatment for TB (streptomycin) was not introduced until the late 1940s. So what did cause the decline? It is tempting to suggest that it was due to improved housing conditions. A reduction in overcrowding would certainly help to limit the spread of TB, and this may have been a factor, at least over part of that period. However, the major advances in this respect were not made until the second half of the 19th century, by which time the level of TB had already fallen substantially.

We cannot rule out the possibility that this was simply evolution in progress. With most infectious diseases, it is not easy to envisage selection of resistance in the host (us, in this case). Our generation times are so much longer than those of the bacterium, so the bacterium will evolve faster than we do. In the case of TB, however, in its heyday it killed so many people of reproductive age (or younger) that it is conceivable that there was enough selective pressure for those who were somewhat less susceptible to be more successful in raising a family. And we do know that there are some human genes that affect our susceptibility to TB.

Although TB is now a relatively uncommon disease in the UK, it is still a massive problem worldwide. It is estimated that there are 9–10 million cases of TB per year, leading to over a million deaths, with the problem being greatest in Africa and in South and East Asia. Furthermore, it is thought that as many as one in three of the world's population are currently infected with the TB bacillus. In order to understand why there are so many more people infected than there are cases, we need to look at bit further at what happens when we breathe in this bacterium.

As we have seen already (Chapter 2), the TB bacillus infects a macrophage in the lungs, and survives and replicates inside that cell. In most cases, our immune system then kicks in and contains the infection, but without eliminating it. So, for every 100 people infected, only five will develop TB within a year of infection. In the remainder, the organism stays within our lungs, contained by our immune system.

As we get older, our immunity starts to wane and the organism then has a chance to develop and spread further. So, of our 100 initially infected (or rather of the 95 who have not developed TB), another five will show the disease at some time later in life. The remaining 90 per cent will never show any signs of the disease (other than the immunological response shown by a skin test, which shows they have been infected).

If we look at people in the UK who have TB, many of those who were born in this country are elderly and are presumed to have been infected many years ago, when TB was more common. A second category are those who have either migrated here from a high-incidence country and have brought the disease with them, or who are the children or grandchildren of such immigrants, and therefore have had more contact with such cases (including possibly frequent visits to countries with a lot of TB).

A personal example: my father had TB in 1948 (although he was probably infected many years before that). This was a less common form, affecting the kidneys. He was one of the first people in the UK to be treated with streptomycin, and he recovered, although not without the loss of one kidney. He used to get cross with me for saying that antibiotics were not responsible for the decline in TB, but it is necessary to distinguish their ability to cure a disease (where they are undoubtedly very important) from their effect on the incidence of TB in the country as a whole. However, although cured of the symptoms, the bacterium was still present and, 50 years later – by which time his immune system would have been much less effective – he developed TB in his other kidney. Fortunately, by then there were even better antibiotics available, so he recovered without further consequences.

There is one other important feature of the TB story. That is HIV/AIDS. If someone becomes infected with HIV, or is already HIV positive before being exposed to TB, then instead of a five per cent chance of developing TB, the chance may be as high as 50 per cent. This applies at a very early stage in the progression from HIV positive to AIDS, so it can happen before there is a major decline in the immune system (usually measured in terms of the count of specific cells, known as CD4 cells, in the blood). This contrasts with other 'opportunist infections', which only start when the CD4 count has fallen to a very low level. So, in the UK and similar countries, amongst the 'indigenous' population, we see cases among those who are HIV positive as well as amongst the elderly.

The advent of effective antiretroviral treatment has greatly reduced this risk – that is, in countries where it is available and affordable. However, this is not the case in many poorer countries, notably in Africa. Here, not only are up to half of the population already infected with TB, but 30–50 per cent of the population may also be HIV positive. The result has been a massive epidemic of TB in Africa. Even if we are sufficiently callous to regard this as someone else's problem, we would still need to be concerned about it, as infectious diseases do not respect international boundaries, and the TB epidemic in Africa poses grave risks to other countries as well.

Leprosy is another ancient disease. Although the diseases described in the Bible as 'leprosy' actually refer to miscellaneous skin conditions, there is no doubt that leprosy has been with us for a very long time. There are many examples of ancient human bones showing the characteristic lesions of leprosy.

The bacterium that causes leprosy, *Mycobacterium leprae*, is related to the tubercle bacillus, and the way it behaves shares some features with it. As with TB, most of the people who come into contact with the disease never develop any symptoms, even though testing for an immune response shows that they have been infected; their immune system deals with it effectively. In those who do develop symptoms, the disease can take several forms. In some cases (the tuberculoid form), the bacteria live within nerve cells and, although bacterial multiplication is well contained (so these patients do not infect others), the immune system damages those cells. This can lead to a loss of feeling in the affected parts of the body – so, for example, a sufferer may not be able to tell that they have picked up something very hot until they smell the burning flesh. Or, they may tread on something sharp, causing a wound in a foot which they do not notice until it becomes severely infected. Since many people in tropical countries, where leprosy now mainly occurs, work barefoot in the fields, this is a severe problem. The subsequent infections and other damage can lead to loss of limbs.

At the other extreme of the spectrum of the disease, the lepromatous form is marked by extensive multiplication of the bacteria in the skin (as *M. leprae* will only grow at temperatures below 37 °C, it only affects superficial tissues). This can lead to extensive and disfiguring lesions, especially on the face, where it may lead to complete destruction of the tissues of the nose in particular. It is this gross disfiguration that is largely responsible for the fear engendered by leprosy, and the consequent ostracization of the unfortunate sufferers, to the extent that the word 'leper' became used in English to mean someone who should be cast out from society (and unfortunately still is occasionally used in this sense, sometimes by people who should know better).

This attitude is extremely unfortunate, to put it mildly, for two reasons. Firstly, ostracism is unnecessary, as there is little danger of catching the disease, since most contacts never develop any symptoms. Secondly, leprosy is easily treated with antibiotics – but only if treatment can start early, before the damage is done.

Anything that deters people from seeking treatment will result in much unnecessary suffering, and the fear of being cast out from society is a powerful deterrent. So, all too often, people hide the symptoms until it is too late.

Although we now tend to regard leprosy as a tropical disease, this was not always the case. Leprosy certainly occurred in temperate regions such as the British Isles in the Middle Ages, and in some old churches it is still possible to see 'leper windows' from which the sufferers could view the proceedings at the altar from outside the church. The last recorded case in the UK (apart from imported cases) was in Shetland in 1798, and it persisted in northern Norway until the last reported case in 1950. Why it declined is unknown, although there is speculation that the decline was linked to the increase in TB.

Turning to another ancient disease, in September 1665, a bundle of cloth arrived in the small village of Eyam (pronounced 'Eem') in Derbyshire. Within a week, the tailor, George Viccars, was dead. Unknown to him, the cloth contained fleas that had brought the plague that had been ravaging London. This set in train a series of events that have made Eyam famous to this day.

By the end of that month, five more people had died; by the end of the year, the death total had reached 42. The rector of Eyam, William Mompesson, concerned that the disease would spread to adjacent villages, persuaded the villagers to enter voluntary quarantine and cut themselves off from the surrounding world. People from neighbouring areas left food for them on a stone at the edge of the village (which can still be seen, Figure 3.2), where coins, suitably disinfected, were left in payment. Eyam suffered terribly, with 260 people, a third of the population, dying of plague by the time of the last death, in November 1666. Whole families were wiped out; for example, Elizabeth Hancock buried her husband and six children within a period of eight days, but did not catch it herself. However, the quarantine measures were successful and the disease did not spread.

The epidemic that devastated Eyam had its origins in London, where the first cases occurred in December 1664. Its arrival had been expected, as there had been 50,000 deaths from plague in Amsterdam in 1663–4. But it was May 1665 before London started to be really alarmed. In June, orders were made, requiring, among other things, that any house in which there was a plague victim should be shut up, with watchmen at the door to prevent anyone, sick or not, from leaving, thus effectively condemning all the inhabitants to death from plague (or starvation).

Over the next few months, there was a mass exodus from London, although those leaving met a rough reception in the surrounding countryside. The economy of London was in ruins, with many dying from starvation in addition to the deaths from plague itself, which reached over 1,000 per week by July. The number of deaths was so great that the bodies had to be unceremoniously tipped into mass graves known as plague pits. The plague continued to ravage London – over 8,000 deaths were recorded in the third week of September – until the autumn, when the epidemic started to decline, although cases were recorded into the early part of 1666. By the end of the epidemic, over 68,000 deaths had been recorded in London (about

Figure 3.2   Eyam Boundary Stone. When the village of Eyam, in Derbyshire, went into voluntary quarantine because of the plague, people from the neighbouring village would leave food on this stone, which was on the boundary between the two parishes. The villagers of Eyam would leave disinfected coins in the holes in the stone.

15 per cent of the population), although the precise figure is unknown. Although London was by far the worst affected, many provincial cities suffered as well.

Two questions arise: why did the plague epidemic start and why did it end? To attempt to answer these, we need to look further at the disease and its cause. Plague is caused by a bacterium, *Yersinia pestis*, which is endemic in rats and other rodents. Diseases that are endemic in animals, and only incidentally infect humans, are known as *zoonoses* – a term I will come back to in Chapter 10. Generally these animals suffer a much milder disease, so it can persist in a rat population where it is spread from one rat to another by fleas. The bacteria multiply uncontrollably in the gut of the flea and block it, so that when the flea tries to feed, it regurgitates the gut content, including large numbers of bacteria, into the wound.

Rat fleas prefer to feed on rats; they will bite humans as well, but only if they are desperate. They cannot survive on humans. However, if the rat population grows beyond a sustainable level, and rats start to die because they can't find enough food, then the fleas leave the dying rats and bite the nearest available mammal – which is often humans. This is one reason for sporadic outbreaks of plague, such as, possibly, the outbreaks that had occurred in London before the great plague, in 1603, 1625, and 1636, which were not inconsiderable (35,000 people died of plague in London in 1625).

One factor that made the 1665 plague worse may have been a population shift. This was shortly after the restoration of the monarchy in Britain, and the large numbers of people who moved to London may have had less immunity to the disease than those who had survived the previous outbreaks. The other reason why we pay so much attention to the 1665 plague is that this was the last time that a major outbreak of plague occurred in these islands. This leads us on to the second question, why did the outbreak stop – and furthermore, why did it not appear again? One popular notion is that the Great Fire of London in 1666 destroyed many of the older buildings, and those that replaced them were built of materials that were less conducive to rat infestation. However, the fire was largely confined to the City of London and did not spread to the areas to the west, which was where the plague started and where many of the deaths occurred. Another explanation that is often put forward is that the black rat was replaced by the brown rat, which tends to live underground rather than in buildings, and carries fleas that are less likely to bite people. However, the brown rat did not arrive in Britain in large numbers until the 1730s. In short, we don't know the answer to this question.

Whatever the answer, there is no doubt that this is a horrible disease. The most characteristic form, known as bubonic plague, occurs when the bacteria localize in lymph nodes near to the original bite. The lymph nodes enlarge and become extremely painful, and they turn black because of haemorrhage under the skin. The disease may arrest at this stage, in which case recovery is possible; however, if it spreads to the blood (septicaemic plague) or to the lungs (pneumonic plague), mortality is 100 per cent if untreated. Only the pneumonic form is directly contagious, being spread by people coughing up aerosols of the bacteria and leading to a much more rapid spread of the disease.

Plague had of, course, been around a long time before 1665. Two epidemics are worth looking at, because of the mark they left on human history.

In 527, Justinian became emperor of what was left of the Roman empire, based in Constantinople (now Istanbul). To the east, he faced the mighty Persian empire. But unknown to him, he also faced a more deadly foe. In 542, plague arrived in Constantinople. We cannot be sure of the precise figure, but a reasonable estimate is that, over a mere three-month period, some 100,000 people in Constantinople died from plague, about one-fifth of the population. Nor was it limited to Constantinople, but it spread throughout the Roman Empire and beyond, returning to Constantinople in 558 and 573.

One of the consequences was a fatal weakening of the Roman Empire. In 636, they faced an Islamic army under Khalid ibn al-Walid at Yarmuk (on the border between modern Syria and Jordan). The defeat of the 'Romans' (actually a miscellaneous collection of soldiers of various origins) at Yarmuk was followed by a rapid expansion of the Islamic/Arab empire throughout the Middle East and North Africa, crossing the Straits of Gibraltar into Spain in 711. Although Constantinople held out (until conquered by the Ottoman Turks in 1453), its power was over, and Europe started on a path to the development of the separate countries that we see today.

For the next major epidemic, we have to move the clock on some 700 years. In June, 1348, the first cases in England of what became known as the Black Death were recorded at Melcombe Regis in Dorset. By September, the disease had reached

London, and over the next 12–18 months it spread over the whole of England and Wales, eventually reaching Ireland and Scotland. By the time the epidemic died out, in September 1350, it is estimated that at least 30 per cent of the population had perished. Subsequent waves occurred in 1361, 1368, and 1371.

One of the consequences was a drastic shortage of labour. England, at that time, was a largely labour-intensive agricultural economy, with a feudal system under the control of large landowners. The shortage of labour forced many landowners to pay higher wages to their labourers, or to rent out their land. Opinions differ as to the extent to which this was a consequence of the Black Death, or whether this was a trend that had already started. But it certainly marks a bend in the road, if not a major turning point, in the development of English society, and especially in ending the feudal system. The Peasants' Revolt in 1381, although not achieving any significant changes in itself, can also be viewed as a signal of the increasing power of the rural working classes.

Although there has been some debate as to whether these earlier epidemics were actually due to plague or some other diseases, this question has now been conclusively resolved, at least for the case of the Black Death. Ancient skeletons have been recovered from 'plague pits', notably one at East Smithfield in London (known to have been used solely for Black Death victims, of whom 2,400 were buried there). Bacterial DNA was recovered by drilling into teeth from some of these bodies (the reason for using teeth being to reduce the extent of contamination by related soil-dwelling bacteria). Using modern technology, the complete genome sequence of this DNA was determined in 2011, and it was shown to be from a strain of *Yersinia pestis* related to all the currently existing strains.

These are just a few examples from a long list of infectious diseases that were much feared killers in former times – typhoid, dysentery, diphtheria, scarlet fever and smallpox are all other examples of diseases that have either disappeared virtually completely – or at least are no longer feared – in the UK and similar countries. The same measures that conquered cholera also put paid to typhoid and dysentery.

Vaccination (which I will look at more fully in Chapter 4) dealt with diphtheria and smallpox – the latter having been eliminated completely, worldwide. Scarlet fever is an odd one; it is still present, but is now such a mild disease that it is hardly recognized. The reasons for this are not known. In the UK, we are now faced with a new set of infectious diseases, labelled as opportunist and emerging infections, which I will look at later in this chapter. But before doing that, we must remember that, for the majority of the world's population, the story is very different. Not only are they still faced with the diseases I have mentioned, but also with a range of other diseases that are still major killers.

Foremost amongst these is malaria (which also used to occur in the UK – see Chapter 10). This is caused, not by a bacterium or a virus, but by a protozoan called *Plasmodium falciparum* (there are several species of *Plasmodium* that cause malaria, but *P. falciparum* is the nastiest). Infection occurs when a human is bitten by a mosquito that has the organism in its salivary glands; the parasite is injected into the victim's blood, in a form known as a sporozoite. In a matter of minutes, these sporozoites find their way to the liver, where they replicate for a short while, emerging in a different form – merozoites. These merozoites infect red blood cells, where they replicate further, before the blood cells break open and release larger numbers of merozoites. This happens in concert, so large numbers of red blood cells break open at the same time, thus explaining the intense fever that occurs at intervals of several days. The released merozoites then go on to infect more red blood cells.

The loss of large numbers of red blood cells causes anaemia, but the worst damage is caused by the infected red blood cells sticking to the walls of the blood vessels and blocking them. If this happens in the blood vessels of the brain, the resulting cerebral malaria is often fatal. This is especially a problem in children, whose blood vessels are smaller and more easily blocked. In Africa, a child dies every 45 seconds due to malaria, mainly for this reason. Altogether, there were 216 million cases of malaria in 2010, with an estimated 655 thousand deaths.

These forms of replication are asexual, but *Plasmodium* also has a sexual stage. Some of the merozoites differentiate into male and female gametocytes. When a mosquito bites an infected person, it will ingest blood containing these gametocytes, which then develop in the gut of the insect into the mature sexual forms, which mate and then go through other stages before eventually leading to sporozoites in the salivary glands, which are then ready to infect someone else.

One of the problems in developing a vaccine against malaria is that these various stages – sporozoites, merozoites, gametocytes (and others I have left out) – have different antigens. So, for example, if you have an immune response against the sporozoites, it will be ineffective against any organisms that have reached the liver and changed into merozoites. And so on. Much effort has gone into trying to create vaccines that combine antigens from different stages, but so far without proven success (although there are candidate vaccines currently in clinical trials).

Effective treatments against malaria are also limited. As we will see in Chapter 4, antibacterial antibiotics rely on the fact that bacteria are different in many respects from human cells. Thus, there are ways to kill bacteria without killing our own cells as well. However, Plasmodia are protozoa, which are eukaryotic and therefore much more similar to human cells, making it more difficult to kill them safely.

There are several drugs that are effective, but not so many – and therein lies a problem. If we use one drug all the time then, sooner or later, the organism will become resistant. In other words, we are applying a very strong selective pressure, so that any mutations that lead to resistance – even if they happen only very rarely – will rapidly take over, and that drug becomes ineffective. I will look further at the drugs available for treating malaria in the next chapter.

3.2 Diseases of the present

So, we have seen that, in the UK and similar countries, many of the diseases that historically were major killers are now diseases of the past – but we should never forget that many of these diseases are still a serious problem for the majority of the world. However, we are still faced with a number of important infectious diseases, which we can consider in three categories:

- diseases which have always been with us, although they may not be historically as important as those already considered;

- 'opportunist' infections, which take advantage of the ability of modern medicine to keep alive people with serious conditions that affect their resistance to infection;

- 'new' diseases (the so-called 'emerging infections') – although, as we will see, some of them are not really new, only newly recognized.

We can consider influenza as an example of a disease under the first heading. During an epidemic, a large number of people die, either of flu itself or, more often, of secondary infections that take advantage of the effect that flu has on our resistance to respiratory infections. However, the number of deaths is usually small in relation to the enormous number of cases, and most people recover quickly. Nevertheless, it is of considerable economic importance, not only because of the cost to our health services, but also because of the number of days of work lost during an epidemic.

The most damaging *pandemic* (world-wide epidemic) on record was that of the so-called 'Spanish' flu of 1918–1919, which killed as many as 50 million people. The mortality rate (the percentage of deaths amongst those infected) was as high as 2.5 per cent, compared to the 0.1 per cent or less that is normally associated with flu. The reasons for this are unclear. Although the 1918 virus is often regarded as a highly dangerous virus, many of the deaths were due to bacterial pneumonia. We also have to consider the unusual factors surrounding this pandemic. The disruptions caused by the war – enormous numbers of refugees, plus the movements of troops at the end of the war – must have contributed to its spread, and the associated stress and malnutrition would have caused a significant drop in natural resistance to infection.

Why do these epidemics occur? We develop a good level of immunity to the virus so, once we have recovered, we do not get infected again with the same virus. But we can get flu again, repeatedly. To understand this, we need to look at the genetic structure of the virus. Firstly, it is an RNA virus, which means that its genetic material is RNA rather than the DNA that is the genetic material of all cells from bacteria to humans. This is relevant, as RNA tends to change more readily than DNA does (for technical reasons that we don't need to go into here). Furthermore, flu virus RNA is single-stranded, rather than double-stranded like DNA. With double-stranded DNA, there are actually two copies of the information. One strand carries the information that is used to code for the proteins that are made, while the other strand (the 'complementary' strand) carries, in effect, a mirror image of it (some genes use one strand, while others use the opposite strand).

The genetic code is made up of four bases – A, T, G and C – and these work in pairs, A opposite T and G opposite C. So, where there is a T on one strand, there will be an A on the complementary strand (don't confuse this with being diploid, which means there are two copies of each chromosome; I'm talking here about the two strands of the same chromosome). Mutations – changes in the sequence of bases – can happen by mistakes being made during the copying of the DNA (or RNA), but if the nucleic acid is double-stranded, there will then be a mismatch at that point – the base on one strand will not be complementary to that on the other strand. Cells have mechanisms for correcting that situation and will therefore correct the new strand so that it does match. This is an important mechanism for ensuring the fidelity of replication; mutations still happen, but at a much lower frequency than might be expected.

But with the flu virus, though, the RNA is single-stranded, so there will be no mismatch detected. This is another reason why it tends to vary quite rapidly. The combined effect means that if someone gets flu one year, they might not be completely immune the following year. Even if the virus is antigenically similar, the antigens may have changed sufficiently to make immunity less than complete. This is known as *antigenic drift*.

However, the major epidemics are due, not to antigenic drift, but to another phenomenon, antigenic *shift*. There are two major antigens that are used to define flu viruses, as these are the ones that our immune system responds to (see Figure 3.3a). They are known as H (haemagglutinin) and N (neuraminidase), and they exist in a number of types which are given numbers. So a specific virus might be referred to as H1N1, or H3N2. One H1N1 virus is not exactly the same as another (remember antigenic drift), but someone who has been infected with an H1N1 virus in the past might have some level of immunity to a subsequent H1N1 strain. They will not, however, have immunity to an H3N2 virus. Thus, when a new antigenic combination emerges, few, if any people will have immunity to it, with the result being a worldwide pandemic.

The 'Spanish' flu of 1918 was an H1N1 strain. More recently, in 1957, there was a pandemic ('Asian' flu) which originated in Southern China before spreading worldwide. This was an H2N2 strain. The next pandemic occurred in 1968 (called 'Hong Kong' flu, as it was first isolated in Hong Kong), and was caused by an H3N2 strain. An apparently highly pathogenic strain of avian influenza virus (H5N1) emerged in Hong Kong in 1997. Although this was brought under control by slaughtering poultry, it re-emerged in 2003, causing infections in birds in many parts of the world, with a limited number of human infections. By May 2009, there had been some 400 confirmed human cases, with 261 deaths. This apparently high mortality rate needs to be regarded critically however, as an unknown number of less severe cases may not have been diagnosed. Fortunately, this virus has not (so far) been able to exhibit sustained human to human infection.

In February 2009, there was an outbreak of a respiratory tract infection in Mexico, although the virus was first isolated in April of that year from two cases in California. It was identified as an H1N1 flu virus, popularly (but misleadingly)

(a) Diagrammatic structure of influenza virus

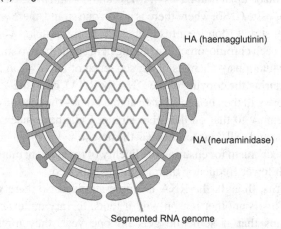

HA (haemagglutinin)

NA (neuraminidase)

Segmented RNA genome

(b) Origins of genome segments in 'swine-origin influenza virus'

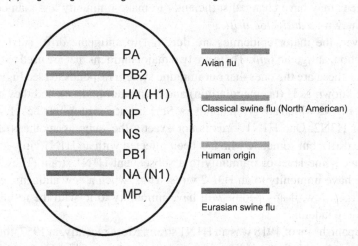

PA
PB2
HA (H1)
NP
NS
PB1
NA (N1)
MP

Avian flu

Classical swine flu (North American)

Human origin

Eurasian swine flu

Figure 3.3 The influenza virus. (a) Diagrammatic structure of the virus, showing the main features referred to in the text. Note that HA (haemagglutinin) and NA (neuraminidase) are further shortened to H and N respectively in virus designations (e.g., H1N1); (b) Origins of genome segments in 'swine-origin influenza virus'. The designations PA, PB2, etc, refer to genes carried by individual segments of the genome. The segments are not actually all the same length.

known as 'swine flu'. Although the initial reports from Mexico indicated a severe disease with high mortality, as it spread to other countries, giving rise to a pandemic, it became apparent that for most people it was a relatively mild infection, typical of most flu outbreaks, apart from people who had some under-lying condition that rendered them more susceptible. Rather unusually, infections were more common amongst younger age groups, indicating, possibly, that older people had some degree of protection through immunological memory of previous H1N1 infections.

The genetic structure of the virus holds the key to these antigenic shifts. In most organisms – even in most viruses – a chromosome contains many genes. The flu virus is not like that. It contains the genetic information for eight proteins, each on a separate RNA molecule. Within the infected cell, they are copied separately and, when the time comes for assembly of the virus particles, one copy of each of the RNA molecules is incorporated. Now consider what might happen if a cell is infected, simultaneously, with two different flu viruses. In the assembly of the virus particles, some of the RNA molecules might originate from one virus and the rest from the other virus. The result is a recombinant virus with some of the properties of each of the two 'parents'.

Animals, especially birds, can be infected by a wide range of flu-related viruses, with different versions of the H and N antigens. Some of these viruses can be devastating to birds, but they are of little direct concern to us because they do not readily infect humans. The risk to humans comes when such a virus recombines with a human virus. Recombination of this sort occurs in animals, especially pigs, which are susceptible to infection with avian (bird-type) and human flu viruses. This is thought to be the major factor leading to the emergence of new pandemic strains at intervals. The so-called 'swine flu' virus provides an interesting example. Its origins are thought to lie in a triple re-assortment (see Figure 3.3b) that occurred in pigs around 1990, between a human virus, a classical pig strain and a bird strain. This strain became established in pigs, with subsequent mixing leading to an H1N2 virus that caused sporadic human infections in USA from 2005 onwards. In parallel with this, an H1N1 bird strain became adapted to pigs (in about 1979), and the final re-assortment occurred when this strain contributed two genes (including N1), resulting in the H1N1 strain that we now know as 'swine flu' – more properly referred to as 'swine-origin influenza virus', or S-OIV.

Another respiratory tract infection, pneumonia, is, among infectious diseases, by far the most common cause of death in countries like the UK and USA. It is especially common in the elderly, where it has long been known as 'the old man's friend', as it can provide a relatively quick release from long-term suffering from other causes. Pneumonia can be caused by a wide variety of microbes, especially when the host defences that normally protect the lungs are compromised. One bacterium, the pneumococcus (*Streptococcus pneumoniae*), although declining in relative importance, still causes perhaps 50 per cent of all cases.

This bacterium has a polysaccharide outer coat, known as a capsule, which is essential for its pathogenicity. Immunity to the capsule can therefore protect against the disease. Unfortunately, there are many strains of pneumococcus, producing different capsules, so a vaccine is used (see Chapter 4) that contains a mixture of several different polysaccharides, thus conferring protection against the most common types (we will also encounter the capsule in Chapter 7, as the ability of extracts from a virulent culture to restore pathogenicity to an avirulent strain led ultimately to establishing that DNA is the genetic material of the cell).

Str. pneumoniae is a versatile bacterium. At one extreme, it can live happily in the throat, causing no problem at all. When it does cause a problem, however, it is not limited to pneumonia. It can also enter the blood stream and spread through it, causing septicaemia, which is usually rapidly fatal. It is also one of the major causes of bacterial meningitis in the elderly and in very young children.

The form of meningitis that causes most publicity is due to a different bacterium, *Neisseria meningitidis* (commonly known as the meningococcus). This is another microbe that commonly lives harmlessly in the throat but which, for reasons unknown, will occasionally break out and cause meningitis and/or septicaemia. Unlike the pneumococcus, it mainly affects older children and young adults – hence the publicity, as this is an age group that is not normally susceptible to life-threatening infections. And life-threatening it certainly is. Although readily treatable with antibiotics in the early stages, it may present as little more than a mild flu-like illness (or a hangover). However, it progresses rapidly, and within 12–24 hours it may be too late. Just going to bed to sleep it off may not be a good idea!

Isolated cases of meningococcal disease occur sporadically and randomly. If there are two or three cases close together, it may indicate an outbreak of a more virulent strain, or it may simply be a chance event. If the apparent cluster of cases is just an unfortunate coincidence, the risk of any further cases is very low – no higher than in the general population – so no intervention is needed. How these possibilities are distinguished, and the control strategies, are considered in the next chapter.

In some parts of the world, though – notably the so-called 'meningitis belt' in sub-Saharan Africa – the situation is quite different and much more serious. There, every December marks the start of the meningitis season, with many thousands of cases. A particularly bad epidemic in 1996–7 saw 250,000 cases, with 25,000 deaths. Epidemics on this scale demand mass vaccination. A new, cheap, vaccine (produced in India) was introduced in 2010 with the aim of vaccinating everyone under 30.

3.3   Opportunist infections

Some microbes are known as 'opportunists', since they only, or primarily, affect individuals whose resistance to infection is particularly low. To an extent, all infections are opportunist, since they will affect some people more than others. However, we usually reserve the term *opportunist* to those infections that only affect a small minority of people – often those who are already in hospital for some other reason. Two organisms that are especially important in this category, and are frequently in the news, are methicillin-resistant *Staphylococcus aureus* (MRSA) and *Clostridium difficile*.

Staph. aureus is a bacterium that is a common cause of minor infections such as boils, and it can cause more serious problems such as wound infections and pneumonia. Once antibiotics became available from the late 1940s onwards, staphylococcal infections were usually readily treatable with penicillin, but a minority of strains produced an enzyme known as penicillinase, or β-lactamase, which is able to destroy penicillin. Around 1960, new derivatives of penicillin (such as methicillin), which were resistant to this enzyme, became available and, hence, these penicillin-resistant strains were sensitive to methicillin. For a short while, this solved the problem, but almost straight away, new strains of *Staph. aureus* started to emerge which were resistant to methicillin. These MRSA strains spread widely in hospitals and caused a major problem of hospital-acquired infections.

Other antibiotics are available, notably vancomycin, for the treatment of such infections. However, of course, the use of vancomycin has led to the emergence of vancomycin-resistant strains, although not yet to a major extent. Infections with MRSA strains, even with those resistant to vancomycin, are usually still treatable with antibiotics. The problem is that it may take a day or two to find out which of the battery of available antibiotics the strain is sensitive to. And if the patient is seriously ill, doctors may not have that sort of time.

The second example is *Clostridium difficile*, which shows the importance of the normal flora in protection against infection. In patients receiving powerful antibiotic therapy, the normal flora of the gut may become disrupted, and this can allow bacteria such as *C. difficile* to flourish. In many cases, withdrawing or changing the antibiotic can resolve the problem but, in some cases, the *C. difficile* infection progresses to a serious condition known as pseudomembranous colitis, which can be fatal even if treated with further antibiotics. Remember, though, that this usually only happens to people who are seriously ill or debilitated to start with.

This is far from the end of the story. Almost any microbe that can grow at temperatures within the body is capable of setting up an infection if provided with a suitable opportunity. Wounds and burns breach the body's first line of defence and can allow infections from a wide range of bacteria. In extreme cases, where a serious accident has exposed brain tissue, there have been examples of bacteria that are normally associated with plant diseases, for example, causing brain infections. Fortunately, this is extremely rare.

However, there are many bacteria capable of taking advantage of even minor wounds, or other defects in our primary defences, which mainly show up as hospital-acquired infections. One common example is the bacterium *Pseudomonas aeruginosa*, which is found in all sorts of moist environments – around sinks and taps, in cloths and mops that are left wet, and so on. A wet dishcloth will quickly acquire the characteristic boiled cabbage smell of *Pseudomonas*. This bacterium is naturally resistant to many antibiotics and to disinfectants – indeed, its metabolic diversity is such that it will actually grow in some disinfectants. Therefore, if a floor mop is kept in a bucket of disinfectant, then the next time it is used, it will spread a culture of *Pseudomonas* over the floor. This is not a very aggressive pathogen, but it can cause serious disease in vulnerable patients, and these may be difficult to treat because of the organism's resistance to many antibiotics.

A rather different cause of hospital-acquired (or hospital-related) infections is the insertion of foreign bodies into the patients. Since this allows microbes to evade many of the defences against infection, it provides an opportunity for infection with a wide range of organisms that are usually unable to invade the body. One of the most common situations – but also the easiest to deal with – is urinary catheterization. The insertion of a catheter into the bladder bypasses the usual ways in which organisms are prevented from reaching the bladder. Once there, they can grow rapidly in the urine. In most cases, this can be dealt with easily by simply removing the catheter. Less easy to deal with are prosthetic devices such as artificial hip joints or heart valves. If these become infected, either during the operation or from the patient's blood stream afterwards, it is a serious matter. Treatment with antibiotics may not be effective, because the surface of these devices is a privileged site for the bacteria. The antibiotics may not reach the site of infection, and the antibodies and immune cells produced by the patient may not, either.

We also see opportunist infections at work in patients with compromised immune systems. This can include transplant patients, where the drugs used to prevent rejection may also lower resistance to infection, or people receiving cancer therapy which can cause reduced immunity. However, the most important cause of immune deficiency leading to opportunist infections is HIV/AIDS (see below).

3.4 'New' diseases

Over the last 30 or 40 years, we have seen a procession of apparently new diseases. Some of these, such as AIDS, do seem to be genuinely new. Others, such as Legionnaire's disease, are either 'new' in the sense that they had not been recognized previously, although they have been around for a long time, or have increased in importance because of the altered conditions of modern day living.

In 1981, the Centers for Disease Control in Atlanta, Georgia, USA noticed a marked rise in the number of prescriptions for pentamidine, a drug used for the treatment of infections caused by an organism known as *Pneumocystis carinii* (at that time classified as a protozoon, but now known to be a fungus). This is an opportunist pathogen, but these infections were happening to people with no known history of a compromised immune system. Further investigation showed an increase in some other opportunist infections among apparently well people, and it was eventually found that there *was* an immune defect – those affected had an abnormally low level of cells that can be recognized by the presence of a specific antigen, CD4. These cells play an important role in helping other cells to respond effectively to an infection, and they are thus known as CD4+ T-helper cells. This enabled a definition of this new condition, as the Acquired Immune Deficiency Syndrome (AIDS). Eventually the cause was tracked down to an infection, when the Human Immunodeficiency Virus (HIV) was isolated by Luc Montaigner at the Institut Pasteur in Paris. Despite what might be claimed in some newspapers, there is now no serious doubt that AIDS is caused by HIV.

One risk factor that was identified at an early stage was homosexuality, as a high proportion of those infected were male homosexuals. We now know that it is a sexually transmitted disease amongst heterosexuals as well as homosexuals. It is also transmitted by the use of contaminated needles, especially amongst intravenous drug abusers, and by transfusion with contaminated blood or blood products.

HIV is a retrovirus; it has an RNA genome but, when it infects a cell, the RNA is copied into DNA, which then integrates into the DNA of the cell. Replication is usually halted at that stage, so someone infected with HIV may carry the virus for years with no ill effects. During this phase, the infected person is HIV positive, but does not have AIDS. However, they can infect others. Eventually, the virus is triggered into replication and, as a consequence, the number of CD4+ cells in the blood starts to fall and the individual becomes subject to a variety of opportunist infections. At an early stage, even before there is a marked fall in CD4+ counts, they become highly susceptible to TB (see earlier in this chapter). The development of TB symptoms in HIV+ individuals is often taken as a marker of progression to AIDS status.

In a country such as the UK, where TB is not so common, patients may escape tuberculosis – but, as their immunity drops further, they become susceptible to other infections. This includes another mycobacterium, *M. avium*, which is not usually pathogenic but is common in AIDS patients, and will multiply extensively in the tissues and the blood. Curiously, it is not clear how much damage this does. Because of the faulty immune system, the patient's body does not react to the bacterium, so they may show few if any symptoms of the infection – it is sometimes said that they die *with M. avium* rather than *because* of it.

AIDS patients can suffer from a wide variety of opportunist infections, ranging from viruses – especially viruses such as cytomegalovirus (CMV), which many people carry as usually harmless latent infections – to fungi (including the yeast *Candida*, which is also a common inhabitant of the body) and protozoa (such as *Toxoplasma* and *Cryptosporidium*).

A major problem with HIV/AIDS, both for treatment and for vaccine development, is the speed with which variants arise. Although there are several effective anti-retroviral drugs available, use of one drug on its own quickly selects for resistant mutants that commonly, and spontaneously, arise during the course of infection in a patient. To counter this, combinations of drugs are usually employed. Similarly, the virus would rapidly circumvent any immunity conferred by a simple vaccine. A variety of approaches have been developed to try to overcome this but, so far, none of the candidate vaccines has proved to be sufficiently effective to warrant large-scale use.

The origins of HIV have been the subject of much speculation. There are related viruses that occur in apes and other primates, and a likely hypothesis is that HIV originated in Africa from one of these viruses, probably as recently as 1950, and for reasons unknown began to spread more widely in the 1970s.

Another new disease was first recognized in 1986, when the first cases were recognized in the UK of a disease in cows, popularly (but misleadingly) known as 'mad cow' disease – misleading, because the cows were rarely 'mad' in the sense of being aggressive, but were more commonly nervous and showed weakness of their legs, leading to difficulty in getting up. Post-mortem examination of the brains of affected animals showed characteristic changes to the structure of the brain, enabling the disease to become more accurately described as 'bovine spongiform encephalopathy' (BSE). A similar disease, scrapie, had been known for a long time in sheep and was first recognized in 1732, and a similar, but rare, condition called Creutzfeldt-Jakob disease (CJD) was known in humans. Transmission of BSE was linked to feeding cows with material derived from other affected cattle (and possibly originally from scrapie-affected sheep). The number of confirmed cases in cattle in the UK reached a peak of some 37,000 in 1992 and declined from then on, with only a handful of cases confirmed in the last few years.

Much public alarm was caused by the occurrence of cases of a human disease resembling, but distinguishable from, CJD, and hence known as 'new variant CJD' (nvCJD), which is believed to arise from eating meat from affected cattle. The first known cases of nvCJD were in 1995 (three deaths), rising to a peak of 28 deaths in 2000 and then declining, although cases still occur occasionally (five deaths from nvCJD in 2011). Although the control measures that were put in place, both to eliminate the disease from cattle and to prevent affected meat reaching the market, seem to have been effective in stopping new infections in humans, cases still occur, due to the long incubation period of the disease (probably up to four years in cattle and longer in humans).

The agent that causes these diseases is a novel one. Purification of the infectious agent shows that it contains no nucleic acid – neither RNA nor DNA – and hence it is not a virus, or any other microbe in the strict sense. The infectious agent, a *prion*, is a protein known as PrP^{Sc}, which is a misfolded form of a protein (PrP^{C}) that occurs naturally in the brain in normal animals. Transmissibility occurs because the abnormally-shaped protein induces a corresponding change in the shape of the protein that is already there. This is a revolutionary concept as it creates an entirely new type of infectious agent. Many scientists still find it difficult to accept, but I find the evidence convincing.

Although classical CJD is not transmissible, it does, to some extent, run in families. The reason is that a mutation in the gene coding for this protein can make it more likely to adopt the pathogenic form spontaneously, without exposure to an abnormal protein from other sources.

The next 'new' disease, Legionnaire's disease, owes its name to the circumstances surrounding its discovery. In 1976, there was a convention of the American Legion (former military personnel) in Philadelphia. Among those attending, there were over 200 cases of an unusual sort of pneumonia, the cause of which was not apparent. There were several possibilities, of which two were most likely. One was that it was an infection with some unknown agent, although the usual tests did not find anything. The second was that it was an allergic reaction to something in the air-conditioning system.

There was therefore a problem in knowing how to treat these patients. If the cause of pneumonia is an infection, then doctors can try antibiotics, although they would not know at this stage if is a bacterium, a virus or something else. On the other hand, if an allergic reaction were the cause, then the answer might be to use something to suppress the immune response, such as steroids. Some clinicians made the latter guess, but it proved to be wrong. This was unfortunate, to say the least, because if a patient has an infection, the last thing that is going to help is to suppress immunity. Thirty-four people died, many of them among the group that had been treated for a supposed allergy. Eventually, the cause was tracked down, although it required the development of new bacterial growth media to be able to cultivate the bacterium, which we now know as *Legionella pneumophila*.

The next question was whether this really was a new disease, or had there been previous outbreaks that had not been recognized? That is not an easy question to find an answer to. How is it possible to tell what caused an unexplained disease many years previously? Fortunately, in some cases, outbreaks had been sufficiently 'interesting' for someone to have the bright idea of collecting serum samples from the patients and storing them away in a freezer, just in case, in years to come, someone might be able to find out what the cause was. One such example had occurred, ironically, in a health department building in Pontiac, Michigan, where there had been an unexplained outbreak of an acute fever, affecting over 100 people, in 1968. When these serum samples were tested, they were found to contain antibodies to *Legionella*, showing that there had, indeed been earlier, outbreaks. Actually, the disease in Pontiac (known as Pontiac fever) had been a milder, flu-like condition.

What causes these outbreaks? We now know that this is a very common organism in the environment, typically in water, but it only causes infections when we are exposed to aerosols containing large numbers of the bacterium. Examples include cooling units (where water is cooled by forcing air through a descending curtain of water), humidifiers in ventilation systems, and showers. One outbreak occurred in a modern hospital in the UK, which had been built with a generous provision of showers for the patients. The first shower cubicle was used frequently and did not cause a problem, but the last in the row was only used when the others were already occupied, and all the cases occurred in patients who had used that shower. During the time when it was not in use, the organism had multiplied within the shower head, so that, when the shower was turned on, the patient was exposed to a massive aerosol of *Legionella*. The immediate answer was simple: they blocked off one of the showers. Having sometimes to wait to use a shower was a small price to pay.

This disease is, to a large extent, associated with modern buildings, and an important part of the prevention of the disease lies in the design and maintenance of these buildings. This includes regular cleaning of water tanks, maintaining hot water systems at a temperature sufficient to kill *Legionella* (even at the risk of scalding some of the patients), and cold water systems at a low enough temperature to prevent (or at least slow down) the growth of the bacterium.

With some of these outbreaks, it is clear that very large numbers of people must have been exposed, yet only a few of them developed the disease. For example, in

2002, a contaminated air-conditioning unit in a municipal arts and leisure complex in Barrow-in-Furness, Cumbria, resulted in a discharge of *Legionella* into the street outside. The consequence was an outbreak of Legionnaire's disease. Although there were 200 cases, with seven deaths, thousands of people must have passed by that building before the outbreak was identified and the cause traced and eliminated. The reason why so few people actually developed the disease is that *Legionella* is another example of an opportunist bacterium. It is much more likely to cause disease in people with certain predisposing factors – which includes smoking, or being elderly or overweight, as well as more obvious factors such as already being ill from some other cause. Hospitals obviously contain a high proportion of people who are ill, which accounts for so many outbreaks of Legionnaire's disease occurring in hospitals.

Given that we know that water is the source of the organism, why can we not just get rid of it by chlorinating our water tanks or disinfecting the pipe work? The problem is that *Legionella* can survive and grow inside protozoal cells within the biofilm that coats the surfaces of tanks and pipes. This protects them against disinfectants, especially when the protozoa form dormant structures known as cysts. The only way to get rid of them is to drain the tanks and scrub the surface to remove the biofilm. I will return to the topic of biofilms and intracellular bacteria in Chapter 8.

I now want to look at another example of an organism that was around for a very long time before its significance was recognized. To understand why, it is necessary to look briefly at how a diagnostic bacteriology lab works. When a specimen arrives in the lab, a small portion is spread onto plates of one or more media. These media are designed to allow the growth of those pathogens that are likely to be present in the specimen and may have caused the symptoms described. No attempt is made to find all the bacteria that may be present. This is not just carelessness, nor cutting corners. With dozens, or maybe hundreds, of such specimens arriving every day, it would not be feasible to try to find everything. Nor would it be sensible. Many of these specimens, such as samples of faeces, will contain hundreds of different bacteria, the vast majority of which are of no significance in relation to the symptoms described. They are our 'normal flora' and can be discounted. So, if no bacterial colonies of significance are found on the chosen media, the specimen is reported as 'No pathogens detected'.

For cases of gastroenteritis, the focus was on detecting *Salmonella* or *Shigella*. However, in 1977, a bacteriologist called Martin Skirrow, working at Worcester Public Health Laboratory, decided that rather too many specimens from people with diarrhoea and vomiting were being reported as 'no pathogens detected' and he decided to do something about it. This involved using different media and different growth conditions, and eventually he succeeded in showing that many of these specimens contained a bacterium he identified as *Campylobacter*. This was not something completely new, as related organisms had been known for many years to cause animal diseases, albeit completely different conditions such as abortion in cattle.

As the isolation procedures pioneered by Skirrow became widely used, it turned out that this was not just some minor contributory factor, but is actually the most common bacterial cause of gastroenteritis, outstripping *Salmonella*. I will say more about food-borne disease in Chapter 5.

I want now to turn to the 'new' bacterium that has had the most revolutionary effect on our thinking about infectious disease – *Helicobacter*, which has the unusual property of living in the stomach. This story begins in 1981 in Perth, Western Australia, where Barry Marshall and Robin Warren were investigating the cause of chronic gastritis.

Traditionally, the stomach has been regarded as essentially sterile, as the low pH is lethal to most bacteria. Occasionally, bacteria can be found in stomach contents, but these are usually just passing through, temporarily protected by being enclosed within food particles. However, Marshall and Warren found, in biopsy material from chronic gastritis patients, an unusual spiral-shaped bacterium.

It was not easy to get acceptance for the revolutionary concept that this condition had a bacterial cause. Could this just be normal flora that nobody had found previously? They tested similar material from other people – the 'healthy controls' – and they found it significantly less often, providing evidence that it is associated with gastritis. That didn't, however, prove that it caused the condition. Two lines of evidence finished the job. Firstly, they looked at the effect of treatment. One sort of treatment suppressed the symptoms but, subsequently, the gastritis tended to recur. The alternative treatment resulted in fewer recurrences, and was found to be antibacterial (not intentionally, but that was the effect). The second piece of evidence came from an unusual, and risky, experiment. Barry Marshall actually drank a culture of *Helicobacter* and, a few days later, became ill with nausea and vomiting. An endoscopy showed evidence of gastritis and the presence of *Helicobacter* in his stomach.

The story doesn't end there. *Helicobacter* was subsequently found to be associated with other diseases, including peptic ulcers, and a form of stomach cancer, and it is now a recognized carcinogen. I imagine that if Marshall had suspected this at the time, he would have been less ready to try his experiment! Marshall and Warren received the Nobel Prize in 2005 for their discovery of *Helicobacter*.

However, it is important to say that the presence of *Helicobacter* in the stomach does not mean you are certain to end up with one of these diseases. Globally, about 50 per cent of people carry it, but most of them show no ill effects at all. A small proportion (maybe as low as one per cent) develop stomach cancer; gastritis and ulcers are more common, but still only a minority (<20 per cent of those with *Helicobacter* in their stomach).

The global figure hides a lot of variation. *Helicobacter* carriage is much more common in developing countries – about 80 per cent in India, for example, compared to less than 30 per cent in the UK. It is also more prevalent in larger families and in people with poor living conditions. It is acquired in childhood, probably from parents or older siblings (although children rarely show any symptoms). Once infected, it is there for life, unless treated with antibiotics. Adults rarely become infected, probably because the adult stomach is more acid and tends to eliminate it before it becomes established by burrowing into the mucus layer that lines the stomach – the mucus layer that protects our own cells lining the stomach from the acid conditions.

I will come back to the subject of microbes causing cancer and other 'non-infectious diseases' in Chapter 10.

3.5   Animal diseases

In Chapter 2, I looked at some of the positive aspects of the normal flora of animals. As with humans, microbes are not all good news for animals, which can suffer from a number of infectious diseases. This can have serious economic consequences. Let's look at just a few examples.

Foot and mouth disease is a highly infectious viral disease that can affect a wide range of animals; in the UK, we are mainly concerned about cattle, sheep and pigs. It is endemic in many parts of the world and it is readily spread by contact and through the air. Although few animals die of the disease, their productivity is greatly reduced.

A serious outbreak occurred in the UK in 2001, originating from a pig farm where it is thought that the pigs had been fed with inadequately treated swill containing meat imported from a country where the disease is endemic. The first cases were detected in February 2001 and, within a few weeks, the disease had spread to many parts of the UK, especially Cumbria and southern Scotland. At the height of the outbreak, 50 new affected farms a day were being reported.

Attempts to control the disease involved culling all animals located near to infected farms – up to 90,000 animals per week – as well as a ban on movement of livestock. This had a severe impact on the farming industry, as well as major costs to the government. In addition, access to affected areas was restricted, including the closure of public footpaths. Since the areas affected included major tourist destinations, there was also a serious effect on the tourist industry.

The final case was reported in September 2001, over seven months after the outbreak started, but restrictions continued into 2002. One reason why this outbreak spread so much faster and further than earlier outbreaks was the extent to which livestock, including infected but undetected animals, were being moved around the country.

One controversy is why vaccination was not used. The reason for this is that export rules would have then prevented the export of British livestock (since vaccination may merely prevent the symptoms rather than the infection, and so vaccinated animals might still harbour the virus) and this would have caused the loss of an economically important trade.

A further outbreak occurred in 2007 in Surrey, and was traced to a site at Pirbright, in Surrey, that housed both the Institute for Animal Health (which was the centre for research and detection of the virus) and a commercial company (Merial Animal Health Limited). A broken underground pipe carrying inadequately disinfected virus had leaked, and from this the virus had been spread to nearby farms. In this case, prompt control measures proved effective, and only a small number of farms were affected.

In 2007, there were well-publicized outbreaks of another cattle disease, Bluetongue. This disease was first described in southern Africa in the 18th century, and for many years was only found in sub-Saharan Africa. Eventually, it reached Europe, where it gradually spread northwards, aided by warmer winters. The significance of the weather is that the disease is spread by biting midges, the prevalence of which is reduced by cold weather in the winter.

In 2007, the disease appeared in the UK for the first time, with 75 farms affected in southern and eastern England. However, in 2008, although the disease remained rampant on mainland Europe, there were no cases in the UK (apart from a few imported animals) – a success story that did not receive any newspaper headlines (good news is no news). This achievement was due to the development of an effective vaccine and its systematic use. High risk areas were identified by computer modelling, based on knowledge of the behaviour and spread of midges. The cooperation of farmers was a vital factor in achieving high levels of immunization of cattle in those high risk areas.

Another success story is the disease Rinderpest (cattle plague), which is caused by a virus related to the measles virus. This is a virulent and rapidly spreading virus, which can kill 80–90 per cent of infected cattle within ten days. It caused serious losses of cattle in Africa, with a series of outbreaks in Nigeria in the 1980s estimated to have caused losses approaching $2 billion. In 1994, the relevant international agencies initiated a campaign to eradicate the disease, using widespread vaccination and long-term monitoring of both cattle and wildlife in affected areas. This has been dramatically successful, and the last known outbreak in cattle was in Kenya in 2001.

However, since the vaccine interferes with tests for the virus, it was not possible to be sure that there were no animals still carrying the virus. In order to confirm eradication, it was necessary to stop vaccination and continue monitoring animals for the presence of the virus. This has now been done, and the UN Food and Agriculture Organization declared (14 October, 2010) that Rinderpest had been eradicated. This makes it the second example (after smallpox) of a disease that has been completely removed from the planet.

It is not just viruses that cause disease in animals. Among bacterial diseases, we can look at bovine TB as an example. We have already considered TB in humans, which is now (in the UK) caused almost entirely by *M. tuberculosis*. However, that was not always the case. Going back to the early part of the 20th century, a substantial proportion of human cases at that time were due to a closely related bacterium, *M. bovis*. If we compare the genome sequences of these bacteria, we see that they are largely identical (so that taxonomically they should be considered as the same species), but there is a major difference in their behaviour. For reasons that are not clear, *M. bovis*, unlike *M. tuberculosis*, is able to infect a very wide variety of animals – probably all mammals (the effects vary widely; some animals, ranging from deer to guinea pigs, suffer an acute disease that progresses rapidly to death, while in others there may be few or no symptoms for quite a long time). So, although we are mainly concerned with cows, due to the economic importance of the disease (and hence refer to it as bovine TB), it is important to remember that other animals also carry the disease – and one of these, the badger, I will come back to shortly.

In the first half of the 20th century, TB was rife in cattle in the UK. At that time, it was not just an economic problem for the farmers, causing loss of productivity, but it was also a cause of human infection, mainly through drinking contaminated milk. Two control measures reduced, and eventually virtually eliminated it as a human disease: tuberculin testing of cattle; and pasteurization of milk.

Tuberculin testing is very similar to the skin test that is used to detect the disease in people. It involves injecting into the skin a small amount of an extract of *M. bovis*. If the cow is infected, it makes an immune response, which shows up as a small lump at the injection site. The infected cows in a herd are then destroyed, and the rest of the herd is re-tested at intervals until all animals are clear of infection. Consumers had the choice of drinking milk from tuberculin-tested (TT) herds, which greatly reduced the incidence of TB, and eventually only milk from herds shown to be TB-free could be sold. At the same time, this scheme gradually brought the disease under control, so that TB in cattle largely disappeared from most of the UK. However, some pockets remained – more about this later.

The other control measure was pasteurization, a process named after Louis Pasteur. This involved heating the milk for a short while to a temperature that was sufficient to destroy the tubercle bacilli (and most other bacteria) without significantly affecting the nutritional qualities of the milk or its taste. This also was an option for consumers – it was possible to buy pasteurized or unpasteurized milk – but it was gradually extended until essentially all cows' milk, and most products from it, were pasteurized.

The combination of tuberculin testing and pasteurization thus meant that milk was no longer a source of TB infection for humans, and the number of *M. bovis* infections has declined to a handful of people per year (many of whom were probably infected many years ago). So why are we still concerned about it?

This comes back to the fact that the testing programme, while greatly reducing the number of infected herds, did not eliminate the disease completely. Pockets of infection remained, especially in south-west England. Furthermore, the disease started to spread again, so that herds in areas previously free of TB developed infection. Although the main factor is likely to be the greater extent of cattle movements that takes place these days, a scapegoat was sought, and this was the badger. In 1971, a dead badger was found on a farm in Gloucestershire where outbreaks of TB in cows had happened frequently. This badger was found to be heavily infected. Subsequent tests on badgers killed on the roads, or on trapped badgers, showed that these animals were quite frequently infected with *M. bovis*, and it was reasoned that the cause of infection in a previously clean herd was contact with infected badgers.

There followed a period in which badgers in infected areas were culled, initially by gassing their setts, and subsequently by trapping and shooting badgers. Here, we need to leave on one side our sympathy for the badger (which actually can be a nasty and vicious animal, rather than the cuddly image we may have of it). After all, especially from a farmer's point of view, there is little logic in accepting the culling of cows while resisting the culling of badgers. The real question is, would culling badgers control the spread of TB in cows?

Initially, this culling programme seemed to work, with fewer outbreaks in areas where the badgers had been eliminated, but subsequent analysis was inconclusive, one factor being that there was a drop in cattle TB in other, non-culled, areas at the same time, reaching its lowest point nationally in 1979. Despite this, there remained a vocal lobby calling for the widespread culling of badgers. Would it work?

To answer this question, a large-scale trial was set up, starting in 1998, in which badgers were culled in some areas in response to detection of TB in cattle in those areas (reactive culling), while in other areas, badgers were killed anyway (proactive culling). In a third set of areas, as a control, no badgers were culled. The results were quite remarkable. Reactive culling was quickly shown to be counter-productive – that is, there was *more* TB in herds in those areas than in the control areas. The reason for this was thought to be that, if badgers are removed from one area, then badgers from the neighbouring areas will move in. This disturbance of the badger population resulted in more contact between badgers from different populations, with the consequent spread of any disease they were carrying.

Proactive culling – killing all badgers in an area whether or not there was TB in cattle – did show a reduction in the trial areas, but there was an increase in surrounding areas, probably from the disturbance of the badger population, as above. Furthermore, the reduction in cattle TB was not great enough to justify the costs involved in trapping and killing badgers, and other measures were deemed more cost-effective. Although TB in cattle costs us, as taxpayers, a lot of money in compensation for the destroyed cows, trying to find and kill all the badgers in even a small area is expensive. This is, of course, not satisfactory to the farmers – nor, indeed, should we be happy with the idea of destroying all these cows. Other solutions are needed. Why not vaccinate the cows?

We do have a vaccine for tuberculosis, which is BCG. Leaving on one side the problem of whether BCG is effective enough (as discussed earlier in this chapter), there is a major problem with vaccinating cows. If a cow has been vaccinated with BCG, it will develop an immune response. So, when a herd is tuberculin tested to determine whether or not they are infected, *all* the vaccinated cows will show a positive reaction, making it impossible to distinguish vaccinated and infected animals.

A lot of research has been devoted to solving this problem, on two fronts: using genetic engineering to develop a vaccine that can be distinguished from an infection, and developing alternative tests that can make this distinction. Although there are still hurdles to be overcome to establish that these measures will be sufficiently effective in practice, we are likely to see them adopted in the near future. Meanwhile, the issue of culling badgers remains a controversial one, with powerful lobbies acting both for and against. At the time of writing (September 2012) the UK government is proposing to allow proactive badger culling in selected areas. How effective this will be, we must wait and see.

The final example, brucellosis (caused by the bacterium *Brucella abortus*), is a highly infectious disease affecting many farm animals, including cattle, sheep and pigs. The infection can remain undetected for a long time, eventually showing itself as abortion. This results in a massive release of the bacteria from the uterus, which causes widespread infection in the herd. Brucellosis can also affect humans, causing a recurrent flu-like illness known as undulant fever, and it can have longer-term consequences such as arthritis and depression. Florence Nightingale is believed to have suffered from brucellosis, which caused her to take to her bed for the latter portion of her life. Infection happens mainly through drinking contaminated milk – a risk that was virtually eliminated by pasteurization.

Britain has been officially brucellosis-free since 1985, although the first year with no new confirmed cases was 1991. The main weapons used to achieve this were vaccination of calves and the testing of herds, with slaughter of any cattle showing a positive test. By 1981, all herds were attested and calf vaccination was stopped. Occasional outbreaks occurred subsequently – in Anglesey in 1993, in Scotland in 2003 and in Cornwall in 2004 – due to importation of affected animals. The effectiveness of the eradication programme makes a contrast to bovine TB; this is probably due to brucellosis not affecting wild animals, and hence the absence of an environmental reservoir of infection.

This is not the end of the story of diseases caused by microbes. In Chapter 5, I will look at food poisoning and food-borne disease and, in Chapter 6, I will consider microbial diseases of plants. But first I want to look at the use of vaccines and antibiotics for the prevention and treatment of infections.

4

Prevention and Cure

In 1969, in a message to Congress, the US Surgeon General, William H Stewart, declared: "It is time to close the book on infectious diseases. The war against pestilence is over." Yet, over 40 years later, such diseases are still with us – so I want now to look at what has been done in the past to counter the threat of infectious diseases and how we can move forward to deal with new and unresolved problems. This includes both preventing or reducing their spread, and treating individual patients if those control measures do not work well enough. Some of the general public health measures have been covered in the previous chapter; the main point here is the use of vaccines and antibiotics. But first, we need to look at the nature of disease spread and, especially, of epidemics.

4.1 Epidemics

The main feature of an active epidemic is that the number of cases in a population increases week by week, or day by day. This does not go on indefinitely. Eventually, the rate of increase starts to slow down and subsequently declines, even if we do nothing at all to check it. The epidemic has run its course. To understand this, we can look at the process in more detail.

Imagine a totally new disease. Nobody has previously had this disease, so we can assume that everyone is susceptible (in practice, the levels of susceptibility would be quite varied, as our resistance to infection depends not just on immunity but many other factors as well – but we can assume, for this argument, that everyone is completely susceptible to this fictitious disease).

Then imagine someone with this disease arriving in a hypothetical country. While they remain infectious, they will pass the disease on to a certain number of people – say, four people. The number of people who catch the disease from an individual case is an important and useful parameter, known as the reproductive rate, designated as R. In the special case when all contacts are susceptible, it is called

---

Understanding Microbes: An Introduction to a Small World, First Edition. Jeremy W. Dale.
© 2013 John Wiley & Sons, Ltd. Published 2013 by John Wiley & Sons, Ltd.

R_0 (the basic reproductive rate). This is, in part, a measure of the infectivity of a pathogen, and it varies from one disease to another. For example, for smallpox, R_0 is quite low (2–4) so 2–4 people would catch the disease. On the other hand, for measles it is high (about 20), so 20 people would get measles from that one initial case. However, it is also affected by other factors, such as the behaviour of the infected individuals and the density of the population – in other words, how many people is the infected case likely to come into contact with while still infectious? If they are very ill and stay at home in bed, they will not pass the disease on to any great extent – or, if they live in the Highlands of Scotland, they will have fewer contacts than if they live in London. But the average values of R_0 for the whole country are still useful.

For our hypothetical disease, we are taking R_0 as 4, so our initial case passes the disease on to four other people. Each of these, in turn, will pass it on to four further people (on average), so we have then 16 new cases, then 64, then 256. If we keep on multiplying by 4 then, in a fairly short time, these will be tens of thousands, then hundreds of thousands of cases. The rate of increase is exponential.

We can then make a further assumption, which is that once someone has had the disease and has recovered, they are immune to it (I am ignoring the possibility that some people might die from the disease). If we consider a situation where a quarter of the population have already had this infection, then, of the people that a single case comes into contact with, only three out of four will be susceptible. So now, instead of passing it on to four people, they will only pass it on to three (again, on average). Therefore, the epidemic starts to slow down.

Mathematically, this is expressed as saying that the value of R, the actual reproductive rate, starts to fall, and is less than the basic reproductive rate R_0. Eventually, we reach a situation where each case only passes the disease on to one other person (R = 1). For this example, that will happen when 75 per cent of the population have already had and recovered from the disease – i.e. of the four people potentially infected, three are immune, so we only get one new case. As each case only gives rise to one new case, the number of new cases stays the same from week to week.

However, as people are still catching the disease and still becoming immune to it, the proportion of the population who are immune continues to increase. The consequence is that now each infected person passes the disease on to fewer than one person. Obviously, it is not possible to have fractions of a person, but if we say that R = 0.9, we mean that, on average, for every ten old cases we get only nine new ones. Now the actual number of cases falls, week by week, until eventually it effectively disappears.

Note that for this to happen, it is not necessary for everyone to have had the disease; not everyone is immune. In this example, the number of cases starts to fall, and will eventually disappear, if 75 per cent of the population are immune. A quarter of the population are still susceptible, but they do not become infected because the chances of contact between an infected person and a susceptible one have fallen below a critical value. This important phenomenon is known as *herd immunity*, and the level of immunity needed to achieve this is called the *herd immunity threshold*.

Some simple mathematics tells us that the level of immunity required to achieve herd immunity is related to the value of the basic reproductive rate R_0. The higher the value of R_0, the more immunity is needed. Mathematically, the proportion remaining susceptible, at the herd immunity threshold, is equal to $1/R_0$. So, if $R_0 = 4$, then the critical value of the proportion remaining susceptible is 1/4 or $0.25 = 25$ per cent. If $R_0 = 10$, then the critical value is $1/10 = 0.1 = 10$ per cent, so herd immunity would not happen until 90 per cent of the population had become immune.

This has important implications for vaccination programmes. Imagine that, instead of letting the epidemic run its course, we step in at an early stage and immunize the population. How many people do we have to vaccinate? In our example, with $R_0 = 4$, we only need to immunize 75 per cent of the population and the epidemic will not happen. However, if $R_0 = 10$, we would need to immunize 90 per cent of the people (the actual targets will be higher than this, as we would have to consider that the vaccine will probably not be 100 per cent effective). This was a crucial factor in smallpox eradication, since the relatively low R_0 value meant that the disease could be contained without having to vaccinate massively high proportions of the people likely to be affected. For measles, with a much higher R_0 value, it is necessary to achieve immunity levels in excess of 95 per cent to achieve herd immunity.

Herd immunity is a very important factor, because it allows protection of those who are not, and cannot, be immunized. This does not just mean those who refuse the vaccine, or those who cannot be vaccinated (e.g. because of immunodeficiency). Populations are not static, and there are new susceptible individuals arriving every day – namely, babies. They are born susceptible, and they usually cannot be vaccinated straight away because their immune systems are not fully developed (this varies from one vaccine to another – some, such as BCG, can be given at birth, but commonly infants are not vaccinated until 12–18 months old). As long as the disease is present in the community, these infants are at risk. But if we can achieve herd immunity, there will be no cases, or at least so few cases that they are unlikely to be exposed to the disease and will therefore be protected.

One further consideration arises from considering the mathematics associated with R_0 values. Some diseases, such as chickenpox and measles, are mainly thought of as childhood diseases. It can be shown mathematically that the average age of infection is related to the R_0 value – the higher the value of R_0, the lower the average age of infection. This makes sense intuitively, as well – the more infectious a disease is, the more likely it is that a child will become infected at an early age. However, if we have a partially successful vaccination campaign (i.e. if we immunize enough people to reduce the level of a disease but without achieving herd immunity, so the disease does not disappear), then the children will not be infected at an early age but, as the disease is still present and they remain susceptible, they may catch it later on.

Mathematically, we have reduced the value of R, but not below 1, so the average age of infection will rise. Instead of the disease being one of childhood, it will affect young (or even older) adults to a significant level. This can have serious consequences for diseases such as measles or mumps, which are more dangerous infections in adults than in young children – and especially for rubella. If we start to get more cases of rubella amongst girls of childbearing age, we would expect to find more congenital abnormalities which are known to be a consequence of rubella infection during pregnancy.

We can now turn our attention to vaccines. The development and testing of vaccines is dealt with in Chapter 9, along with some further examples, so for now I will mainly look at the general concepts and how vaccines are used. The currently used vaccines are of three types: live vaccines; killed vaccines; and subunit vaccines (including toxoids).

Live (attenuated) vaccines consist of a strain of the pathogen which has lost its virulence, so that it is no longer able to cause disease, but it still retains its immunogenicity. In other words, it will still cause an immune response which will protect against infection by the real pathogen.

The first vaccine to be used was of this sort. In centuries past, there was a long-standing belief that dairymaids were especially pretty, and this had a basis in terms of resistance to smallpox. A related disease, cowpox, was prevalent in cattle, and it is thought that the close association that dairymaids had with the cows led to them being infected with cowpox. This is a much less virulent disease in humans than smallpox, but the viruses are similar and those who got cowpox did not subsequently get smallpox. Since cowpox does not lead to the disfiguration associated with healed smallpox lesions, this explained the supposed beauty of the dairymaids.

In 1796, Edward Jenner attempted to exploit this phenomenon by injecting some material from a cowpox lesion into the skin of an eight-year old boy, James Phipps, and he subsequently did a challenge experiment by deliberately attempting to infect the boy with smallpox. This was a highly risky experiment that we would not consider nowadays but, fortunately (for the boy and for Jenner), the boy survived.

There was a considerable amount of controversy surrounding the use of this vaccine, not to say ridicule (see Figure 4.1). Nonetheless, such was the fear of smallpox that the practice of vaccination caught on, and it formed the basis for modern vaccine technology; indeed, the very word 'vaccine' owes its origins to this source (Latin *vacca* = cow). The vaccine that was a key component of the strategy that successfully eliminated smallpox (see later) was a related virus known as vaccinia.

Another important live vaccine is the oral polio vaccine (the Sabin vaccine, which may be familiar as the 'sugar-lump' vaccine – one way of administering it was as a drop of liquid applied to a lump of sugar to take away the taste). In the first half of the 20th century, polio was a common and serious disease of children, leading, in some cases, to severe and permanent wasting of muscles and lifelong disabilities, In extreme cases, the muscle wasting was so severe that the patients had to be kept alive with artificial respirators –the cumbersome machines known as 'iron lungs'.

The Cow Pock — or — the Wonderful Effects of the New Inoculation! — Vide. the Publications of ƴ Anti Vaccine Society.

Figure 4.1 *The Cow Pock or the Wonderful Effects of the New Inoculation!* James Gillray (1756–1815). Jenner is about to vaccinate a woman sitting in a chair, while other patients show the effects of the vaccine, in the form of cows sprouting from different parts of their bodies. (*Source*: US National Library of Medicine)

The introduction of polio vaccines, however, has made this a disease of the past, to the extent that, throughout most of the world, polio has been eradicated. Complete eradication is a real possibility, and it would have been achieved by now if it were not for the resistance to vaccination in a limited number of countries, for a mixture of political and religious reasons. The situation is improving, and there are now fewer than 1,000 cases per year worldwide – mainly in Pakistan and Afghanistan, and in several countries in Central and West Africa.

Although the Sabin polio vaccine has proved to be immensely successful in controlling polio, it also demonstrates one of the potential downsides to a live vaccine. It tends to revert. In other words, further mutations can occur that restore virulence. This is not a problem for the person vaccinated because, by the time reversion occurs, they are immune and therefore do not suffer any problems. However, the vaccine virus does multiply in the body for a while, and is excreted. It can therefore infect others who have not been vaccinated.

In the early days of polio vaccination, this was not a significant problem. When there was a lot of polio around, a few rare cases due to transmission of a reverted vaccine strain were not significant and, indeed, would not have been noticed. But as the vaccination campaign became more and more successful, and naturally transmitted polio became less common, release of the virulent virus from a few people who had been immunized started to become the main reason for the persistence of the disease. Therefore, the use of the live vaccine was discontinued and present attempts to stamp the disease out in the few remaining pockets relies on a killed vaccine (the Salk vaccine). This was actually the first to be introduced, but it has to be given by injection and was therefore less popular – hence the predominant use of the live (oral) vaccine in the initial stages.

One further example of a live, attenuated, vaccine is the TB vaccine BCG, which has already been considered in Chapter 3.

Killed vaccines are generally much easier to produce, by growing the pathogen on a large scale and then treating it, chemically or by heat, to inactivate it. Of course, the product has to be tested thoroughly to make sure that no live pathogen is present. Unfortunately, killed vaccines tend to be much less effective, requiring booster doses to achieve long-lasting protection. At least in part, this is thought to be due to the ability of a live vaccine to persist in the tissues, while a killed vaccine is eliminated from the system.

There are few vaccines of this type now in widespread use. The best example is the influenza vaccine, where there is a need to respond rapidly to the emergence of new pandemic strains, which makes a killed vaccine useful. The killed polio vaccine (Salk vaccine) was referred to above, and there are a handful of other viral vaccines (such as the rabies vaccine) for specific purposes. Many of the older bacterial vaccines, such as those for cholera and typhoid fever, were killed vaccines, but they suffered from a major disadvantage, in that that the immunity they produced was not very strong and did not last long. In general, the killed vaccines have been supplanted, either by live vaccines (as is the case with the typhoid vaccine) or by subunit vaccines.

A subunit vaccine, as you might guess, consists of parts of the pathogen rather than the whole thing. The rationale is that sometimes (especially with bacterial vaccines), even if the bacteria are dead, some of their constituents can be toxic. Therefore, it is better to use purified fractions of the cell extract and so remove any toxic products.

One example is the vaccine against pertussis (whooping cough). This is a respiratory tract infection caused by the bacterium *Bordetella pertussis*, and is an unpleasant disease, occurring mainly in children. It is marked by paroxysms of violent coughing, followed by a sudden intake of breath (the 'whoop'). Although most children recover successfully, there is a significant level of neurological dis-orders amongst the survivors. The bacterium owes its pathogenicity, in part, to a combination of protein toxins that affect cells in the lungs. These toxins have to be removed from the vaccine, and the vaccine in use currently is an acellular (subunit) vaccine that consists of fractionated cell extracts from cultures of *B. pertussis*.

Pertussis vaccination is highly successful, with the number of cases in the UK falling markedly following its introduction in 1958 – but it has been controversial in the past. In the 1970s, there were (unjustified) fears over the safety of this vaccine. As a consequence, the uptake of the vaccine declined and fell below the level needed to achieve herd immunity. This led to epidemics of pertussis. In the winter of 1978–79, there were over 100,000 cases in the UK, resulting in many deaths and even more cases of severe neurological problems. I will come back to the question of vaccine safety and how it is assessed in Chapters 9 and 10, including considering the more recent controversy regarding the use of the MMR vaccine.

Where the symptoms of the infection are due to a single toxic protein, we can adopt a different approach. Rather than removing the toxin, it is possible simply to purify and inactivate it, and to use that as the vaccine. This is known as a *toxoid*. The inactivated toxin is completely safe but it retains its antigenic properties, so it will produce a protective antibody effect with the result that, when someone encounters that pathogen, their antibodies will neutralize any toxin that it produces. Two of our most widely used, and most safe and effective, vaccines come into this category – diphtheria and tetanus. These vaccines are considered further in Chapter 9.

The advent of modern genetic manipulation techniques have opened up new routes for the production of subunit vaccines. Instead of separating components from the whole pathogen, the gene(s) that code for a specific protein, or proteins, can be cloned and expressed in a safe and easy to use host. One example, that of the hepatitis B vaccine, is described in Chapter 9. Another example is the human papilloma virus vaccine (used primarily to protect girls against cervical cancer), where expression of the relevant gene in a recombinant host is used to produce the viral coat protein, which spontaneously assembles into virus-like particles which are highly immunogenic, but completely safe as they contain none of the viral DNA.

This approach has enormous advantages, especially in terms of safety. The fact that the pathogen itself is not involved in the vaccine production process completely rules out any possibility of the vaccine causing disease, as well as removing the hazards of growing a pathogenic microbe on a large scale. Additional applications of gene cloning, in the genetic modification of pathogens to make specifically attenuated strains, and in making novel live vaccines by introducing other genes into a characterized vaccine vector, are also described in Chapter 9.

Having got a vaccine, how should we use it? Mass vaccination, simply vaccinating as many people as we can persuade to be vaccinated is the most obvious, but not always the best, strategy.

If the disease is common, then mass vaccination probably is appropriate, but we still have to consider what uptake we need (assuming we are aiming to achieve herd immunity), and whether this is achievable. Cost-effectiveness is also a factor. If the vaccination campaign is successful, the level of the disease will decline so much that it is no longer worth trying to vaccinate everybody. A good example is the eradication of smallpox.

Once mass vaccination had reduced the level of smallpox to a considerable extent, this approach meant that a lot of resources were being put into immunizing people who would never get the disease anyway. The alternative strategy then was to use the vaccine in response to outbreaks. In the UK, there was an outbreak of smallpox in 1962 in South Wales, and many people (myself included) were immunized as part of a successful attempt to control that outbreak. A similar strategy was used to mop up the remaining pockets of smallpox in other parts of the world. Fortunately, smallpox is an easily recognized disease, so it was possible to train individuals in remote communities to recognize cases of smallpox. The response team could then be alerted and move in to vaccinate everyone who might have been in contact with the infected person. Another factor that was key to this strategy is that smallpox is not infectious until the symptoms appear. The same strategy would not work with measles, for example, where the early symptoms are fairly nondescript, and someone who is infected can pass the disease on at this stage. By the time the recognizable rash occurs, the damage has been done.

Meningitis vaccination usually needs a more subtle approach. The most dangerous form of meningitis is due to the bacterium *Neisseria meningitidis* (see Chapter 3), which is unusual in that it can kill healthy young adults very quickly. Because the disease is not common, mass vaccination would not be appropriate but, if an outbreak occurs (i.e. there are several cases in a community), then vaccination of contacts in that community can be used to prevent further spread. However, this is only relevant if there is a genuine outbreak – that is, the cases are due to infection from the same source. When there are several cases, there is often a lot of public pressure for vaccination to be used, but it may not be a real outbreak. The disease occurs sporadically, in a random fashion, and it is quite possible for several cases to happen close together without any transmission having occurred between them. This is simply a chance event. How can we tell whether it is an outbreak or not? The organisms can be tested to find out if they are the same strain or not. If they are the same strain, then it may be an outbreak (although there is still the possibility of chance infections with the same strain). If the strains are different, then it is not an outbreak, so there is no point in vaccination.

An alternative way of targeting the use of a vaccine is through identifying, and immunizing, those people who are at most risk from the disease. The most obvious example is people who are travelling abroad; there is no value in being vaccinated against yellow fever, for example, unless travelling to an area where yellow fever is endemic. The meningitis vaccine can also be used for groups of people who are at special risk. It is commonly offered to university students, for example.

One group of people who have a high level of risk for a variety of diseases are the elderly, both because of a higher likelihood of infection (due to a declining immune system) and because the consequences are more likely to be severe. Thus, influenza vaccination is normally offered to people over 65, as they are more likely to have serious consequences following influenza. Similarly, pneumonia (and other infections) due to the bacterium *Streptococcus pneumoniae* is a major cause of death in the elderly, and the risk can be reduced by immunizing with a pneumococcal vaccine.

The above description of the immunological basis of the use of vaccines envisages the body making an immune response to a protein antigen. However, many bacteria have an outer coat called a capsule, which is made up of poly-saccharides (large molecules consisting of chains of sugars joined together – see Appendix 1). This tends to shield any bacterial proteins (or other antigens) so that antibodies cannot gain access to them. An effective immune response in such cases, therefore, has to be directed towards the polysaccharides in the capsule.

Haemophilus influenzae is one example. This bacterium, despite its name, does not cause influenza but is often associated with other respiratory tract infections. More seriously, it can also cause meningitis, especially in young children under five years old. It is possible to make a vaccine from the polysaccharide capsule, but there is a problem. Polysaccharides do not produce a very good immune response, especially in young children whose immune system is not fully developed – and this is the very age group that needs protection. The way around this is to link the polysaccharide chemically to a protein, such as the tetanus toxoid. This *conjugated* vaccine is much more effective, and it is the basis for the currently used Hib vaccine.

We also need to think about antigenic diversity. For some pathogens, there are several strains which have different antigens, so a vaccine prepared using one strain will not necessarily protect against a different version. We have already seen one example of this, with influenza, where the vaccine has to be made using the virus that is currently in circulation, but in some cases there are several different strains permanently in circulation. With the polio virus, for example, there are three antigenic versions (three *serotypes*), so the vaccine has had to contain a mixture of all three types. Now, however, in the last stages of the eradication campaign (hopefully), there is often only one strain remaining, so it is possible to use a monovalent vaccine.

With *Strep. pneumoniae* (pneumococcus), it is not so easy. The key antigen here is the polysaccharide capsule, as with *H. influenzae* but, with the pneumococcus, there is a very large number of different serotypes (probably more than 100). It is clearly impossible to use a vaccine containing so many different antigens—but all is not lost. Not all of those serotypes are equally common, and most of the cases are due to one of a handful of types. Therefore, the current vaccine contains the antigen from a selection of the most common types. This does cause further potential problems, however. The serotypes in circulation in the UK are different from those in other countries, so a UK vaccination would not give adequate protection to someone travelling abroad. Furthermore, it is to be expected that, if the vaccine provides effective immunity against the selected strains, the other strains will be at an advantage and we may therefore see a shift in the distribution over time, which means the vaccine would have to change.

One of the most remarkable examples of antigenic diversity comes from the protozoan parasite that causes sleeping sickness (*Trypanosoma brucei*). When you are infected, you develop an acute fever as the protozoa multiply in the blood, but this soon subsides as you make antibodies that virtually eliminate the parasite. Virtually, but not completely. A small minority escape, because they change the key antigen on their surface, so the immune response does not recognize them. These multiply and cause a second bout of fever, which in turn subsides as you make a new set of antibodies. This cycle recurs again and again, as the parasite works its way through a very large set of potential antigens (over 1,000 of them). How it does this is covered further in Chapter 7. The overall message here is that the antigenic diversity of this parasite makes it extremely difficult to develop a vaccine against sleeping sickness.

There are a number of other diseases, especially those that are prevalent in many tropical countries, where it has proved very difficult to produce an effective vaccine. For malaria, for example, although there have been several candidate vaccines that have shown promise in clinical trials, none of them are as yet effective enough to be worth using. Vaccines for HIV/AIDS have also yielded disappointing results in clinical trials. And control of tuberculosis would be greatly helped by an effective vaccine, as BCG does not seem to offer sufficient (if any) protection.

In my more pessimistic moods, I wonder if a vaccine against TB is a wild goose chase. In the earlier account of TB, I said that only ten per cent of those infected ever develop symptoms (apart from those infected with HIV). The rest mount an effective immune response even without a vaccine. So maybe those who get the disease are, for some reason, unable to produce the required immune response. We also need to consider that, in areas with a very high incidence of TB, it is quite common for someone who recovers from the disease to subsequently become re-infected—with a different strain, so this is not just reactivation of the earlier infection. The implication is that, for these individuals at least, actually having the disease does not establish protective immunity, and this is bad news for vaccine development.

The final message is that, although vaccines have been immensely successful in combating a range of infectious diseases (and continue to do so), they may not always prove to be the ultimate answer.

4.2   Antibiotics

In 1928, Alexander Fleming made his now famous observation that the growth of staphylococcal colonies on an agar plate was inhibited around a fungal colony that

Figure 4.2   Sir Alexander Fleming. (Reproduced by permission of Wellcome Library, London)

had contaminated the plate. The fungus was a species of *Penicillium*, a common mould that produces spores that are distributed in the air, and which can easily contaminate plates in a laboratory. The inhibition was subsequently shown to be due to a chemical produced by the fungus, which became known as penicillin.

Although the story of antibiotics usually starts with the discovery of penicillin, this was not the first therapeutically useful agent, even discounting those highly toxic and non-specific chemicals, such as arsenic compounds, that had been used previously. In the 1930s, sulphonamides were synthesized as an accidental by-product of the chemical industry – the key discovery being that the dye Prontosil Red, made by the German company IG Farbenindustrie, was able to control streptococcal infections in mice. The active component of the dye was shown to be sulphonamide. Gerhard Domagk was nominated for a Nobel Prize in 1939 for this discovery, but he was not allowed to receive it by the Nazis (he did receive the medal in 1947). Although this was a German discovery, related compounds were soon produced in many other countries, and it may have had a decisive impact on the Second World War; Churchill developed pneumonia in 1943 and recovered after being treated with a sulphonamide (developed by the British company May and Baker).

The main difference between penicillin and the sulphonamides was that sulphonamides were made chemically, while penicillin is a natural product. For a long time, the term 'antibiotic' was reserved for natural products of microbes, so sulphonamides were not regarded as antibiotics, but as synthetic chemotherapeutic agents. This distinction is no longer useful (if it ever was), especially as some antibiotics that were originally discovered as natural products are now made chemically, while many others are chemical modifications of natural products. So I will refer to all of these agents as antibiotics.

The story, of course, does not end with penicillin. The significance of penicillin sparked off extensive searches for similarly useful compounds produced by other microbes. This yielded a number of important antibiotics, including streptomycin, chloramphenicol and tetracycline. Most of these additional antibiotics were not from fungi (as penicillin was) but from various species of a group of filamentous bacteria, *Streptomyces*, that live in the soil and are amazingly prolific when it comes to producing antibiotics. The story of the development and production of antibiotics is considered further in Chapter 9.

The advent of antibiotics had a dramatic effect on the treatment of infectious diseases. Death from previously feared diseases became, almost overnight, the exception rather than the rule. Although only limited amounts of penicillin were available during the Second World War, it was invaluable in limiting the damage caused by wound infections. No single antibiotic was capable of treating all possible infections, and one of the main reasons for needing to search out new compounds was to extend the range of diseases that could be cured. Streptomycin, for example, was important as the first really useful treatment for TB (for which penicillin was no use).

But all was not plain sailing. Penicillin is a very safe antibiotic, with only minor side-effects (apart from a dangerous immune response in a tiny proportion of people), but that was not true for some of the subsequent discoveries. Streptomycin, for example, could cause deafness. Similarly, chloramphenicol and tetracycline had problems with toxicity, and the use of these antibiotic is now limited; they have been supplanted, for most purposes, by newer and safer antibiotics.

Resistance was a further problem. When penicillin was first introduced, most isolates (about 98 per cent) of *Staph. aureus* were sensitive and were readily killed by it. However, a small proportion was not and, as the use of penicillin increased, these resistant strains became more prevalent, until within a few years most strains in hospitals were resistant to it. This resistance was due to the production by the resistant strains of an enzyme (penicillinase) that destroyed the antibiotic. Eventually, methods were developed (see Chapter 9) to produce semi-synthetic modified penicillins, one of which – methicillin – was shown (in 1960) to be resistant to penicillinase attack and, hence, able to kill *Staph. aureus* strains that were resistant to penicillin. However, within a year from the introduction of methicillin, new strains of *Staph. aureus* were found that had developed resistance to methicillin (see Chapter 7). These methicillin-resistant *Staph. aureus* strains are the now familiar MRSA that cause such a problem in hospitals.

This battle between the introduction of new drugs and the capability of bacteria to evolve resistance to them has occurred with other bacteria and other antibiotics, although there are exceptions. Another feared pathogen, *Streptococcus pyogenes*, which causes severe wound infections and septicaemia ('blood poisoning'), as well as a variety of other diseases such as scarlet fever, has remained universally sensitive to penicillin. The reason for this remains a mystery.

We now need to look more closely at how antibiotics work, and how bacteria develop resistance. I will then come back to considering the future prospects for antibiotics.

For an antibiotic to be useful, it has to exhibit what is called 'selective toxicity' – that is, it has to kill the bacteria (or at least stop them growing) without killing us at the same time. This doesn't have to be absolute. As long as the concentration that will affect the bacteria is much lower than those concentrations that are damaging to the patient, the antibiotic can be useful. How do they do this?

The most obvious way is that they affect a target that is essential for the bacterium but is not present (or at least not essential) in human cells. For example, penicillin prevents the final step in the synthesis of a structure, peptidoglycan, that is an essential part of the cell wall of most bacteria. Without this structure, the cells simply burst. Mammalian cells do not have peptidoglycan – indeed, they do not have a rigid cell wall at all – and so they are unaffected by penicillin.

Sulphonamides are a second example of this sort of target site specificity, as they prevent the synthesis of the vitamin folic acid. All cells need folic acid; most bacteria make it for themselves; if they are prevented from making it, they will not be able to grow. Human cells, on the other hand, do not make folic acid, so are not sensitive to the effects of sulphonamides. These drugs do show some toxicity, but that is due to another factor; if too much is present, it is deposited in the kidneys.

However, many antibiotics, including streptomycin, chloramphenicol and tetracycline, affect protein synthesis, and the mechanisms of protein synthesis in bacteria and human cells are basically similar. With these drugs, the selective toxicity mainly relies on a different factor. For the antibiotic to affect protein synthesis, it has to get into the cell. The antibiotic is then selectively toxic, because it cannot get into a human cell, or at least not at a level at which it would have a damaging effect.

Turning to antibiotic resistance, there is a lot of doom and gloom around, with some people prophesying the end of the 'antibiotic era'. Much of this is overblown, but there is something of a problem, so it is worth looking at how bacteria (or indeed other microbes) develop resistance to antibiotics.

The simplest way that a microbe can become resistant to an antibiotic is by mutation of the microbial gene coding for the protein that is the target for the antibiotic. A small change in the nucleic acid, often a single base, results in a slightly different protein with a different amino acid at one point in the chain. This protein must be important, otherwise inhibition of it by the antibiotic would not affect the microbe. Thus, the vast majority of the many such changes that could happen will actually be lethal – the bacterium, or virus, will not survive or will not be able to replicate. Some other changes, although they would not be lethal, will not affect the sensitivity of that protein to the antibiotic, and so they will not cause resistance. However, a tiny minority of these changes are such that they still allow the enzyme to work (although possibly not as effectively as the wild type), but will also reduce or abolish the ability of the antibiotic to bind to the protein. Therefore, the bacterium becomes resistant to the antibiotic.

Bacteria have another trick up their sleeve. Many bacteria carry, in addition to their main chromosome, one or more smaller bits of DNA known as plasmids. These plasmids can, in some species, be transferred from one cell to another – very often to other species. The plasmids can carry genes that are responsible for a wide range of different properties, but the most notorious are genes for resistance to antibiotics. In fact, it was the ability of antibiotic resistance genes to move from one bacterium to another that originally led to the recognition of bacterial plasmids.

Antibiotic resistance genes on plasmids typically work in a different way from the resistance mutations described above. Plasmid genes often code for enzymes that destroy the antibiotic, or modify it in such a way as to make it inactive. The penicillinase that is responsible for the penicillin resistance of many *Staph. aureus* strains is located on such a plasmid. Although, in general, these genes only cause resistance to a single drug (or to drugs belonging to one chemically similar group), a plasmid can carry many such genes, and can therefore cause resistance simultaneously to several antibiotics, including drugs other than the one the infection is being treated with. So in this case, treatment with, say, penicillin, might provide a selective pressure in favour of a strain carrying a plasmid with genes for resistance to chloramphenicol and streptomycin, as well as penicillin, and the infecting organism becomes resistant to drugs that it has not encountered.

The ability of plasmids to transfer between species has another important consequence. A person may be infected with a completely sensitive bacterium, but they carry, in parts of their body such as the mouth, throat, or intestines, vast numbers of bacteria of many different types – and some of these may well carry plasmids that confer resistance to several different antibiotics. Those bacteria are doing no harm, but the plasmids that they carry may be transferable to the infecting organism. This will only happen rarely, but when someone is treated with anti-biotics, again the selective pressure can select for those rare events, and the antibiotic will no longer be effective. I will come back to plasmids in Chapter 7.

Let's think about mutations a bit further. Bacteria do not 'do it on purpose'. All sorts of mutations happen all the time, due to mistakes in copying the DNA. Most of these mutations are likely to be damaging. However, exposure to an antibiotic represents a new environment, and hence a new opportunity for specific mutants to thrive. The frequency with which a specific mutation happens, randomly, is extremely low – often in the region of one in 10^9 cells for a typical bacterium, or one in a thousand million. In the absence of the antibiotic, such mutations are of no benefit to the microbe, and the mutant will not survive. Once the antibiotic starts to be used, though, there is a very powerful selective pressure operating. Any resistant mutants are now able to multiply and will have the opportunity to take over, with the consequent spread of drug resistance.

Although a mutation frequency of one in 10^9 cells sounds very low, it is readily detectable in the lab if we remember the enormous number of bacteria present in a culture. *E. coli*, for example, will readily grow to 10^9 cells per ml. If an antibiotic such as streptomycin is added to such a culture, it will quickly kill off almost all the bacteria, but there will be a few that contain the mutation required for them to survive. It is a similar story with an infection. The numbers of bacteria (or viruses) present in the body may well be large enough for it to be quite possible that, even before the patient starts taking an antibiotic, there are already a few mutants present that will make the microbe resistant.

One way of preventing this, or at least reducing the risk very considerably, is to use more than one drug. If the frequency of mutation to resistance to either drug by itself is 1 in 10^9, then the chance of mutation to both simultaneously is (1 in 10^9) times (1 in 10^9), or 1 in 10^{18}, which is vanishingly small in relation to the nature of infection. So, for example, because HIV can mutate very readily, AIDS patients are usually treated with a combination of antiviral agents. Similarly, TB patients, who need lengthy antibiotic treatment (thus carrying a greater risk of selection of drug resistant mutants) are also treated with several anti-TB drugs. If resistance is due to a plasmid, this strategy may not work, since the plasmid may confer resistance to several antibiotics at the same time. Fortunately, this does not happen with TB, as *M. tuberculosis* is one of the few bacteria that does not seem to carry plasmids naturally.

The origin of plasmids is a different matter. Here, rather than random spontaneous mutations, a bacterium becomes resistant by acquiring a pre-formed genetic element that already carries genes conferring resistance to one or more antibiotics. Where do these plasmids come from? Chapter 7 covers plasmids more fully, but for the moment it is sufficient to be aware that plasmids of one sort or another occur in most species of bacteria, in a proportion of strains. They can move around between strains, or between species, and can acquire new genes from their host, so building up a repertoire of genetic information. If they happen to be of use to a strain in a specific environment, then that strain will flourish and the plasmid will also flourish. In the presence of antibiotics, this means plasmids that happen to carry resistance genes will be selected for.

That doesn't fully answer the question, of course. Where do the resistance genes come from? That is a more difficult question, and one that has not been conclusively proved. However, the most likely scenario, and one that is generally believed, rests with bacteria in the soil – and especially with the *Streptomyces* that produce many naturally occurring antibiotics. How can a bacterium make antibiotics and not be killed by them? One way is to have evolved not only to make the antibiotic, but also to become resistant to it. Similarly, other bacteria in the soil may have developed resistance to counter the effect of the antibiotic produced by the *Streptomyces* that they live with. Plasmids in these organisms may then acquire resistance genes, and the plasmids can then be passed from one bacterium to another, eventually ending up in a pathogen and causing us a problem.

So, the use of antibiotics is a potent driving force behind the spread of drug resistance – but far worse is the misuse of these drugs. If you go to the doctor with a cold, there is usually no point in being given antibiotics, as they will have no effect on the virus that causes the cold. Similarly, if you have gastroenteritis, it will probably get better by itself; antibiotics may not help much. Occasionally, their use may be justified in such cases. For example, if you are especially vulnerable, there may be a serious risk of a bacterial infection following a cold – or, the gastroenteritis may be so severe that therapy is needed. Also, the diagnosis may be so uncertain that an antibiotic is needed to cover other possibilities. However, in too many cases it is simply that the drug is prescribed because the patient is not satisfied with just being told to go away and wait for it to get better. The predictable outcome is that any bacteria the patient is harbouring (and, as we have seen, there are enormous numbers of them) are likely to become resistant to that drug.

This is more of a problem in hospitals, not only because there are so many more opportunities for resistant bacteria to be passed from one patient to another, and because there are so many highly susceptible patients, but also because a hospital is likely to use a much wider variety of antibiotics and, hence, there are opportunities for the development of multiple drug resistance. A good hospital will have a policy in place for restricting the use of antibiotics, rotating the commonly used ones and keeping specific drugs in reserve for patients who really need them. This, of course, needs to be combined with effective measures for limiting the extent of cross-infection between patients (or from staff and visitors to patients).

Inadequate treatment is also a problem. People who have been prescribed antibiotics tend to stop taking them as soon as they feel better, rather than completing the course. Not only does this incur the risk of a relapse, as the infection may not have been completely eradicated, but it also encourages the development of drug resistance. It may be argued that effective therapy, which would kill all but the rare resistant mutants, would provide a stronger selective pressure than incomplete treatment, which would also allow some sensitive ones to survive; and, indeed, there is some apparent logic to that argument. Nevertheless, inadequate treatment does seem to be a factor in promoting microbial drug resistance.

The situation is worse in many countries where antibiotics can be bought over the counter, so that self-medication is common and, often, inappropriate. Furthermore, as the antibiotics may be relatively expensive, patients may stop taking the antibiotics early and sell the left-over drugs to someone else. Problems of antibiotic resistance tend to be much worse in such situations.

We also have to bear in mind that antibiotics are also widely used in farm animals. Using antibiotics to treat a cow with mastitis is not a problem – the milk is discarded until there is no trace of the drug present. However, using antibiotics prophylactically (e.g. giving drugs to a whole herd or flock to prevent infection) is much more likely to result in the development of resistant strains. Even more controversial is the use of antibiotics as growth promoters. For reasons that are not completely understood, giving antibiotics continuously at levels below those used for treatment can increase the rate of growth of the animal.

There are regulations in place in various countries to attempt to control the veterinary use of antibiotics, by discouraging their use as growth promoters (prohibited in the EU) and by limiting the drugs used prophylactically to those that have no human use. There is a problem with the latter point, though. Antibiotics come in families, so if a drug that is used only in animals is related to one that is used in humans, there may be selection for resistance to both. For example, the drug avoparcin used to be used as a growth promoter in animals, and had no human use. However, the use of avoparcin was accompanied by increased occurrence in animals of bacteria resistant to the related drug vancomycin – and vancomycin is the main drug used for treatment of MRSA infections. This use of avoparcin was banned in the EU in 1997.

Analyzing such a situation requires foretelling the future. Virginiamycin was, at one time, widely used in animals and was thought to be safe, as there were no related drugs used in humans. However, new drugs belonging to this class of antibiotics (streptogramins) were subsequently found to be valuable human antibiotics. It is important to put this in perspective. The extent to which antibiotic resistant bacteria in animals are a source of resistance in human pathogens is controversial. What we can be sure of is that the misuse of these drugs in humans is a much more important factor.

As I said earlier, the spread of antibiotic resistance has led many commentators to prophesy the end of the antibiotic era. We hear frequently that some 'superbug' has become resistant to all known antibiotics, and that infections will soon become untreatable. Such claims need to be treated with more than just a grain of salt. The overwhelming majority of bacterial infections *can* be readily treated with antibiotics, and are likely to remain so. In which case, why do people die from these infections?

There are two problems. The first is that antibiotics by themselves do not cure infections. It is the patient's own defences that are mainly responsible for eliminating the pathogen, and the antibiotics buy time for this to happen. They reduce the numbers of bacteria and slow down their growth, so that the patient's defences can act to finish the job. So, if a patient has an impaired immune system – someone with AIDS, for example, or a transplant patient on immunosuppressive therapy to prevent rejection – then even the most aggressive antibiotic therapy may not be successful. The same is true for patients with less obvious impairments to their immune system – the elderly, in particular. It also applies to infections in sites that the body's defences cannot reach effectively, including prosthetic devices such as artificial hip joints or heart valves. If these sites become colonized by bacteria, it can be very difficult to treat this effectively, because of the absence of help from the immune system. One of the consequences of medical advances is an increase in the numbers of people in all of these categories.

The second problem is that it can take time to find out which antibiotic is most likely to work. Even with a rapidly growing bacterium, it can take 24 hours to isolate the microbe and a further 24 hours to find out which is the best antibiotic. A seriously ill patient may well be dead before their doctors have found out which antibiotic to use. In these circumstances, the doctors have to start off by guessing what the bacterium is and what it is likely to be susceptible to. Obviously, they have a lot of experience to go on in reaching this decision but, if the bacterium has become resistant to a drug that is usually effective, then the guess will be wrong and the patient may be lost before an answer is found.

So far, I have mainly been considering antibiotics that work against bacteria. What about other pathogens? First of all, there are fungi and protozoa. These are eukaryotes (i.e. they have a nucleus, in contrast to bacteria), and therefore they are fundamentally much more similar to human cells (which are also eukaryotic) than are the prokaryotic bacteria. It is thus much more difficult to devise ways of killing fungi and protozoa without having toxic effects on the patient. Consequently, there exists only a very limited repertoire of antifungal or antiprotozoal drugs, and some of those have serious problems of toxicity.

The most notorious of these is melarsoprol, a drug used in the treatment of late-stage sleeping sickness (due to the protozoan *Trypanosoma brucei*), where some estimates put death from the effects of the drug as high as 30 per cent (although more sober estimates suggest 5–10 per cent). There are less toxic drugs available for treating the early stages of the disease but, by the time the disease has invaded the brain, these drugs do not work. Melarsoprol is used because, once the patient is in a coma, they would die anyway if untreated – so the chance is worth taking. It should also be said that there is another drug that could be used at this stage, but it is much more expensive and may, therefore, simply be unaffordable in the poor countries where this disease is endemic.

There is, of course, the major exception of malaria, caused by various species of *Plasmodium* (which are protozoa), where an effective drug has been available for a very long time. Long before the European invasion of South America, the inhabitants were using an extract from the bark of the cinchona tree to treat fever, and we now know the active ingredient of this as quinine. During the 20th century, limitations to the supply of quinine (exacerbated by the two world wars) led to a search for chemically synthesized substitutes. The main outcome was chloroquine, which became the mainstay of antimalarial therapy until its effectiveness was compromised by the development of resistance. Other drugs followed on, but resistance appeared to those as well. The latest addition is another natural product, artemesinin, derived from a traditional Chinese treatment using extract from the plant wormwood (*Artemisia annua*). There are already signs of resistance appearing to this as well.

At the other end of the microbial scale we have viruses, which are only able to replicate inside host cells, and use much of the machinery of the host cell to achieve this. It is very difficult to prevent the replication of the virus without, at the same time, affecting the host cell. Thus, the antibiotics that are so useful for treating bacterial infections are useless against viruses. Nor can lack of penetration into the cell to achieve selective toxicity be relied upon, because the drug needs to get into the cell to have any effect on the virus. Therefore, the only possible targets are the limited range of activities that are specific to the virus.

For the simpler viruses, there are very few of these – for example, the influenza virus has only eight genes, so an effective anti-flu drug has to target one of the activities encoded by one of those genes. One such drug is Tamiflu (oseltamivir) which blocks the action of the viral neuraminidase, thus stopping it from spreading from one cell to another. In contrast, herpes viruses, such as those that cause cold sores (as well as more serious conditions), are more complex and have some 80–100 genes, which makes it rather easier to find a target (but still much harder than finding good antibacterial agents). One drug available for herpes infections is acyclovir, which is an analogue of one of the bases that make up DNA. A viral enzyme converts it into an active form, so that replication of the DNA of the virus is prevented. The host cell does not have an enzyme that is able to activate the drug, so uninfected host cells are (relatively) unaffected.

Detailed knowledge of the genetic make-up of a virus, and hence of the proteins it codes for, makes it possible (but not easy) to design tailor-made drugs that will target virus-specific enzymes. However, going from a chemical that will inhibit a specific viral enzyme to having a drug that will be clinically useful is not straightforward. Even so, there have been notable successes – in particular, the battery of antiretroviral agents that have had such a marked impact on the death rate from AIDS. For example, there is one class of drugs designed to inhibit the enzyme that copies the RNA of the virus into DNA. Our own cells do not do this, so they do not have a susceptible enzyme, and they are therefore not affected.

Viruses, of course, can also develop resistance, usually by mutation of the target site that makes it insusceptible to the action of the drug. HIV, for example, mutates very readily to become drug-resistant. The best way to counter this is to use several drugs simultaneously, as mentioned above. The chance of the virus mutating to become resistant to, say, three different drugs simultaneously, is infinitesimal.

5
Microbes and Food – Friend and Foe

In previous chapters, we saw how bacteria can help to prevent infection as well as causing it. When it comes to the food we eat, the many-sided behaviour of microbes becomes even more marked. Microbes in the food can make it go off, or they can make us ill. On the other hand, we use microbes extensively to preserve and produce food. In this chapter, I will look at some examples of these many facets of microbes in relation to the food we eat and drink.

5.1 Food spoilage

The concept of food 'going off' is a familiar one. Most fresh food will become spoiled, in one way or another, if kept too long or under the wrong conditions. It does not necessarily mean that it would make someone ill if they ate it; it would just be unpalatable. Let's look at some examples of how it happens.

Raw milk from a healthy cow, taken under clean conditions, contains low numbers of bacteria (10^2–10^3 per ml). However, milk is an excellent medium for growth of many bacteria and, if not kept cold, any organisms that are in the milk to start with (as well as others that contaminate the milk subsequently) will multiply and cause the milk to 'go off'. Traditionally, milk went sour, which was due to acid production by lactic acid bacteria fermenting the lactose in the milk. This is now rarely seen, partly because pasteurization reduces the numbers of such bacteria, and partly because of refrigeration, since these bacteria will not grow at the low temperatures in a fridge.

More commonly seen nowadays is the phenomenon known as 'bitty cream' – the milk may look good and may have not much of a smell but, when it is added to a hot drink, it forms particles on the surface. This is due mainly to *Bacillus* bacteria (which are very common all around us, and hence can readily

Understanding Microbes: An Introduction to a Small World, First Edition. Jeremy W. Dale.
© 2013 John Wiley & Sons, Ltd. Published 2013 by John Wiley & Sons, Ltd.

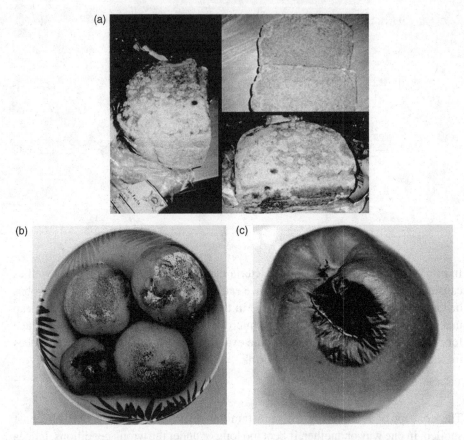

Figure 5.1 Food spoilage. (a) Mouldy bread (*Source:* United States Envrionmental Protection Agency); (b, c) rotten fruit.

contaminate the milk). These bacteria form spores which can survive the heat treatment of pasteurization, and some strains can grow, albeit slowly, at refrigeration temperatures. They make an enzyme called lecithinase, which alters some of the fat in the milk producing small globules of fat and protein that clump when added to a hot drink. This is not likely to make anyone ill, but it also doesn't look or taste good.

Pasteurization of milk was originally introduced over 100 years ago, mainly as a way of prolonging shelf-life. The fact that it also eliminated milk as a source of some human diseases was incidental. Originally, it involved heating the milk to 61.7 °C for 30 minutes, but nowadays a higher temperature (71.7 °C) and a shorter time (15 seconds) are used (the odd numbers arise because the temperatures were originally specified in °F). This combines minimal alteration to the taste and texture of the milk with effective elimination of the relevant bacteria. Although these conditions fall well short of those needed to get rid of spore-forming bacteria, there does not seen to be a risk from the really nasty bacteria, such as *Clostridium botulinum* (which is covered later in this chapter).

Milk keeps even better if heated to higher temperatures (135 °C for 1 second for UHT milk, or over 100 °C for 20–40 minutes for sterilized milk). The disadvantage is that the properties of the milk are changed; sterilized milk in particular can have a rather nasty taste, and is less often used nowadays.

For our second example, consider meat and fish. Fresh meat inevitably becomes contaminated with a variety of microbes during processing. If it is kept cold, most of these microbes will not grow – but some can. Initially, this will not be noticeable, as the bacteria will start by using the carbohydrates (such as glycogen and glucose) that are present in the meat. These changes are not obvious. However, these easily used substrates soon get used up, at which point the bacteria switch to the major component of the meat, proteins and their breakdown products, amino acids. Bacteria growing on amino acids make a wide range of products, including sulphur-containing compounds such as hydrogen sulphide (H_2S), and these are responsible for the smell of the meat as it goes off. One of the major groups of bacteria responsible, the pseudomonads, also commonly produce slime, which adds to the spoilage of the meat. Pseudomonads will only grow in the presence of air, so vacuum-packed meat keeps longer. It will go off eventually, though, due to other bacteria which grow more slowly. These tend to produce acids, which impart a sour smell to the meat.

Fish tend to contain much less carbohydrate, so the contaminating bacteria start to use amino acids almost immediately, which is why fish goes off much more quickly. Some fish, such as dogfish and shark, contain high levels of urea. Bacterial breakdown of urea produces ammonia very rapidly, which not only makes the fish uneatable but also can contaminate other fish stored nearby; therefore, in many areas, fishermen tend to discard these fish when they catch them. Shellfish such as lobster have very high levels of free amino acids and tend to go off more rapidly than fish in general, which is why they are kept alive until just before they are eaten.

Finally, let's look at fruit and vegetables, which are rather different micro-biologically. Fruit are mainly too acidic for common environmental bacteria to flourish. Spoilage of fruit is therefore dominated by fungi, and especially moulds. The blue or green moulds that appear on the surface of many fruit are typically members of the *Penicillium* genus, which are found everywhere in the air. These are related to the microbe from which penicillin was first discovered. The colour comes from the spores that they make on the surface. These spores are liberated in great numbers, so one mouldy orange can quickly contaminate a whole batch. One of these moulds, which causes spoilage of apples, is especially significant because it produces a toxin (patulin) which can cause problems in unfermented apple juice.

Vegetables are, in general, less acidic, and many bacteria (as well as some fungi) can cause spoilage. Some of these are plant pathogens, so they may cause disease before harvest as well as spoilage subsequently. These range from the bacterial rots and scabs of potatoes to a species of *Botrytis* (a fungus) that causes neck rot of onions. With both vegetables and fruit, any breaks in the surface, whether by bruising or by insect damage, will allow opportunist microbes from the air to colonize and cause spoilage.

Among other plant products, cereals (such as rice, wheat, maize and millet) are especially important, as they constitute the major part of the human diet for most of the world. Cereals are relatively dry when stored properly and, as such, are not a very good place for microbes to grow. The trouble mainly comes when they are not stored properly, resulting in mould growth. This is not a small problem. Some of these moulds, notably members of the *Aspergillus* genus, produce powerful toxins known as mycotoxins.

Similar problems occur with nuts, and especially groundnuts (peanuts). The mycotoxin produced by *Aspergillus* that is of special concern is aflatoxin. This was first recognized in 1959, following an outbreak in East Anglia, when several thousand turkey poults died due to contaminated groundnut meal used as a protein supplement. Aflatoxin is not only an acute toxin, but is also one of the most carcinogenic substances known (at least in experimental animals). If people can afford to throw away their mouldy corn, then it doesn't matter. However, this is often the major food source – in some cases virtually the only food source – for many people living on the verge of subsistence. If the alternative is starvation, then they will take a chance on it. The result, in some cases, is a severe outbreak of disease. In 1974, 1,000 people were affected in India through eating mouldy maize; about 100 of them died. In 2004, an outbreak in Kenya caused 125 deaths out of 317 people affected.

5.2   Food preservation

The most obvious way of stopping (or at least slowing) food spoilage is to keep it cold. However, refrigeration and freezing have only become a routine possibility relatively recently, and they are still not available to many people across the world. Hence, throughout the centuries, people have developed other ways of preserving food. This is especially important where food is only available intermittently and we rely on being able to keep that food to see us through the winter months,

As we saw in the previous section, it is sometimes possible to reduce the numbers of microbes, without seriously altering the properties of the food, by mild heating (pasteurization). But, although this may increase the shelf life to some extent, the effect is limited; pasteurized milk still goes off quite quickly. Furthermore, while pasteurization does destroy some dangerous pathogens, it is far from sufficient to eliminate all the risk. In particular, it will not destroy bacterial spores (such as those of *Bacillus*, or the more dangerous *Clostridium* species, which includes those causing botulism). Even boiling does not usually kill these. For this, higher temperatures are needed, such as those achieved in a pressure cooker – at a pressure of 15 pounds per square inch (in old units), a temperature of 121 °C can be reached. This is enough to kill bacterial spores, but often at a cost of unpalatable changes to the taste and texture of the food.

Thus, there is a dilemma. Most ways of preserving foods (discounting modern methods such as freezing and freeze-drying), if they are to be effective, change the properties of the food. The traditional way around this dilemma is to accept it and to regard the preserved food as a desirable product in its own right.

The most commonly used methods involve drying (and smoking). Dried fish and meat are common components of the diet in many cultures, kippers being perhaps the most familiar example in the UK. In addition, many foods can be preserved, not by drying in the common sense, but by making the water they contain unavailable to any microbes in the food. Microbiologists use a concept called water activity to define this process. The water activity of food can be reduced by adding sugar, which will be familiar to anyone who makes their own jam. As is so often the case, nature showed us the way here – honey keeps virtually indefinitely. Jam and honey are far from dry in the normal sense (honey has up to 20 per cent water), but the high level of sugar they contain makes the water unavailable for microbial growth. In general, fungi such as yeasts and moulds are able to tolerate lower levels of water activity than bacteria, so jam may develop mould on the surface, or yeasts may grow within it.

Honey has some interesting additional properties. The nectar collected by the bees often contains a variety of contaminating microbes, and it takes days for the bees to transform it into honey. As the temperature within the hive is 33–35 °C, what prevents these contaminating bacteria from growing? In part, this is due to beneficial bacteria carried by the bees in their honeycrop – including lactobacilli and bifidobacteria, which are related to the 'friendly bacteria' in yoghurt. Some honeys (especially manuka honey – named from the manuka tree in New Zealand) also contain antimicrobial agents such as methylglyoxal.

Also, honey has traditionally been used for treatment of wounds, and still is in many parts of the world. It is said that honey was used to treat a face wound suffered by Prince Harry in the battle of Shrewsbury in 1403. However, proper clinical trials of using honey to treat wounds have been scarce, so how useful it really is remains controversial. I should add that, in general, the table honey that is sold in shops has low antibacterial activity, so I do not recommend using it for treating wounds.

Salt can also be added to foods to lower the water activity, and is often used in combination with drying and smoking for the preservation of fish and meat – for example, in the traditional processes of curing bacon and ham. Salt has antibacterial action in addition to its effect on water activity, and hence is a very valuable preservative. Historically, it was an important part of the economy and it has left its traces in the language – the word 'salary' is derived from salt, and its importance is seen in many place names around the world, such as Salzburg in Austria, and Nantwich in Cheshire, England (although the connection may not be obvious unless you look it up).

Nowadays, many other preservatives are used. Nitrite is especially important, because of its ability to inhibit spore-forming bacteria such as *Clostridium botulinum*. This is not without some controversy, though, since nitrite can react with other compounds to produce nitrosamines, which are carcinogenic. Although the levels in food are quite low, awareness of the potential problem has resulted in reductions of the nitrite levels used – but a balance is necessary to avoid an increased risk of botulism.

An alternative process involves making the food acidic, which is familiar as pickling (as in pickled onions). This can be done simply by adding acid (such as vinegar), or the same end can be achieved by microbial fermentation. The process of fermentation, which extends beyond merely preserving the food, warrants a separate section.

5.3   Fermented foods

Fermented foods are very much a part of everyday life throughout the world. We would think straight away of fermented drinks such as beer, cider and wine. To these we can add bread, and also fermented milk products such as yoghurt and cheese, as well as a wide variety of other fermented products ranging from fermented vegetables, meat and fish, to soy sauce and vinegar.

Before looking at some of these processes, a reminder of what we mean by 'fermentation' (as introduced in Chapter 1). Many organisms, ourselves included, 'burn' our food using oxygen. The most efficient way of doing this involves a chain of cytochromes, with electrons passed from one to another until eventually we get to oxygen. This is known as aerobic respiration. Many (but not all) bacteria do much the same thing, although some of them can carry out respiration in the absence of air, but still using cytochromes (anaerobic respiration). Fermentation, in the strict sense, is different in that it does not involve a chain of cytochromes. Instead, the cell harvests what energy it can from the reactions that convert one intermediate to another within the cell. For example, in alcoholic fermentation, energy is obtained from several of the steps in the conversion of glucose into ethanol. Since oxygen is not involved in this process, it can be carried out in the absence of air. This is a much less efficient way of using fuel than respiration (see also Appendix 1 for further information).

Let us start by considering alcoholic fermentation by yeasts. Other microbes can produce alcohol as well, but it is yeasts (especially the yeast known as *Saccharomyces cerevisiae*) that we normally employ to produce drinks such as beer and wine. Although yeasts can carry out typical aerobic respiration in the presence of air, once the supply of oxygen becomes limiting, they will switch to fermentation and the production of ethanol. Anyone who has tried brewing their own beer will argue that they have not done anything to exclude air, but that does not matter. The yeasts will start to grow aerobically but, by the time they have got going, they will use up the oxygen faster than it can diffuse into the beer, so they will start to make ethanol instead.

Bread is a bit different. Here, the process is aerobic (so not strictly a fermentation by the above definition) and the yeast will convert the sugar all the way to carbon dioxide, causing the dough to rise.

For many other fermented products, the fermentation uses bacteria rather than yeasts. Bacteria have much more diverse fermentation pathways and can produce a wide range of different products. Some of these may be what is wanted – mainly acetic and lactic acids – depending on the product. Others will contribute to the final flavour of the product, 'nice' or 'nasty', depending on what bacteria are in there, as well as varying according to the conditions used.

If you drink half a bottle of wine and leave the rest for another day, you may well be disappointed by the result. Bacteria can get into the wine and convert the ethanol into acetic acid, giving a sour taste (the word vinegar comes from the French *vin aigre*, meaning 'sour wine'). Traditional vinegar production starts with a process very similar to making an alcoholic drink. Bacteria, such as *Acetobacter*, are then added and the mixture is incubated with aeration, since air is necessary for the bacteria to convert the ethanol into acetic acid. Acetic acid can also be produced chemically, and if you buy cheap 'vinegar' you may notice that it is labelled as 'non-fermented condiment'.

Another obvious use of fermentation to extend shelf life is in milk and dairy products, where the use of fermentation probably dates back to the first use of milk from animals – sheep and goats to start with (about 9000 BCE), with cows being domesticated from about 6000 BCE. Initially, the fermentation would have happened through the action of bacteria that occurred by chance in the milk, which must have produced some quite unreliable results. Nowadays, specific bacteria are added to ensure a consistent product.

Yoghurt is made by the combined action of two bacteria, *Streptococcus thermophilus* and *Lactobacillus bulgaricus*. The former tends to predominate in the early stages as it grows faster, but the lactic acid that it makes from the lactose in the milk causes the pH to drop (i.e. it becomes acidic) and the *Lactobacillus* comes into its own, as it is happier than the streptococcus at low pH. As the pH drops, the proteins in the milk coagulate to form the familiar gel structure. The low pH of the final product (3.8–4.2) makes it unfavourable for growth of other bacteria, so it keeps well, and it does not usually cause food poisoning. However the addition of other ingredients such as fruit and nuts to the fermented product can change the picture. In 1989, there were cases of botulism in the UK associated with hazelnut yoghurt.

Another important way of preserving milk, with a long history, is to convert it into cheese. Some 14 million tonnes of cheese are produced per year, worldwide. The initial process is similar in principle to that for yogurt (although different bacteria are used, depending on the sort of cheese). The main differences come in the subsequent processing. For a hard cheese such as cheddar, the curds formed by the initial fermentation are compressed to expel the whey and the cheese is then left at a low temperature to mature. Bacterial enzymes continue to act during maturation, and these give the cheese its characteristic flavour. For some cheeses, such as Stilton, the maturation is assisted by internal growth of moulds (in this case a species of *Penicillium*), which produce the characteristic blue veins.

There are many other fermented milk products, including kefir, which is made extensively in Russia and other countries of the former Soviet Union, and koumiss, a drink made from fermented mare's milk in eastern Europe and central Asia. Buttermilk is now commonly made by the fermentation of skimmed milk rather than taking the liquid that separates during the churning of butter.

Vegetables can also be preserved by the production of acid during fermentation. Most familiar, at least to a European audience, is sauerkraut. Here, the cabbage is often fermented by naturally occurring microbes, rather than by adding specific ones, and the unwanted microbes are inhibited by adding salt to the cabbage at the start of the process. One virtue of sauerkraut is that vitamin C (ascorbic acid) is partially conserved. Sauerkraut was used extensively for the prevention of scurvy in the Dutch navy in the eighteenth century, and Captain Cook gave his sailors two servings of a pound a week during his voyage of 1772. Other pickled fermented products familiar in the West include olives and pickled cucumbers/gherkins, while in Asia, a wide variety of vegetables are preserved in this way.

Amongst fermented meats, the most widespread are fermented sausages, of which there are an enormous variety, including salami, chorizo and pepperoni. The characteristic features of the various products are caused by a combination of different meats, different starter bacteria (sometimes moulds, too) and various fermentation conditions, as well as the addition of salt and spices. As well as their keeping quality, the safety of the product must be considered; it is essential to prevent *Clostridium botulinum* (which causes botulism) from growing in the sausages (the word 'botulism' is derived from the Latin for 'sausage'). This is achieved by the production of sufficient acid within the sausage as well as by adding salt; sodium nitrite is often added as an additional preservative.

Some fermented foods, particularly tempeh, derive a considerable part of their nutritional value from the microbes (moulds in the case of tempeh) that they contain. It is only a short step from that to using the moulds themselves as food. They can easily be grown on a large scale, using a simple growth medium (mainly glucose and ammonium salts). Since moulds grow naturally as filaments, they can easily be harvested from the medium and woven into a product (mycoprotein) with an acceptable texture and a high protein content. The main example of such a product, marketed as Quorn, is made from a mould (*Fusarium*) grown in large fermenters.

There are some much more familiar examples of using microbes as food. The most obvious is yeast extract, which is produced from the yeast residues from a fermentation process. Less obviously microbial foods are mushrooms and similar fungi. Although these are clearly not microscopic, they are simply the fruiting bodies produced from a mass of underground mycelia, which are long chains of more or less independent cells. As argued in Chapter 1, it is the existence of these microscopic cells that provide the rational basis for including fungi in general as microbes.

If we are to include mushrooms as microbial food, we should not forget algae (seaweed). This is not commonly used in Britain – although we should not forget the traditional Welsh delicacy laverbread (*bara lawr*) – but seaweeds are widely used in the Far East (e.g. Japan, China, Korea).

5.4   Food poisoning and food-borne diseases

Now we come to the other side of the coin. We have seen how microbes can be used to preserve, and even make, food, as well as how they can spoil it. But they can also make people ill – in some cases, severely.

In 1983, I went to Calcutta to help to run a course in genetic manipulation, then a relatively new set of techniques. I was expected at the Guest House in the early evening, so they had left a cold supper waiting for me in my room. However, my flight from Bombay was delayed by several hours and I didn't get there until midnight, by which time my supper had been sitting in a warm room all evening. I had not had much to eat all day and was very hungry. I looked at this cold chicken leg and, in the end, I couldn't resist it. Mistake. The next morning, diarrhoea set in, and for two days I barely had the strength to crawl from my bed to the toilet and back again. It was my own fault – I should have resisted the temptation.

This is a typical food poisoning story, with bacteria such as *Salmonella*. Chickens, in particular, are commonly contaminated with *Salmonella*, especially when reared intensively. Added to this is the frequent cross-contamination during transport and processing. If they are adequately cooked, most, if not all, of the *Salmonella* are killed, so eating freshly, and thoroughly, cooked chicken is unlikely to cause a problem. The infective dose is usually quite high; in other words, we can eat small numbers of *Salmonella* and not become ill. Food poisoning usually happens if the bacteria have grown in the chicken (or other food), because it has been kept out of the fridge for several hours. Remember that these bacteria grow quickly, with an optimum doubling time (at 37 °C) of some 20 minutes. Even if the conditions are not quite that favourable, we might expect them to increase tenfold every two hours or so. If we start off with, say, 100 bacteria (which is a very small number in bacterial terms), after six hours we would have 100,000 bacteria.

Although that is the conventional story, it is not always like that. In 1982, there were 245 reported cases of infection with an unusual *Salmonella* strain, with isolated cases spread over the whole of the UK. This was unusual, as food poisoning usually happens in local clusters. Intensive investigation tracked the source down to bars of chocolate that were contaminated with a very few *Salmonella* (which cannot grow in bars of chocolate). It is possible that the fat in the chocolate protected the bacteria from the stomach acid, so the bacteria were able to reach the intestines, which means that just a few bacteria in the chocolate were able to make someone ill. In another instance, in 2006, Cadbury recalled a million bars of chocolate after a leaking pipe contaminated a production line – for which they were fined £1 million. Although the firm said the contamination was at a low level, there is no 'safe level' for the number of *Salmonella* in chocolate.

This raises a question: how can a manufacturer know that a product is completely free of *Salmonella* (or anything else)? Usually, a small sample is tested, but if one chocolate bar out of a thousand has one bacterium in it, the only way it can be found is by testing the whole of every bar – which leaves nothing left to sell. The recommended alternative is to use HACCP (Hazard Analysis Critical Control Point) analysis, which basically means identifying and monitoring the critical risk points in the process, thus ensuring that the process itself is safe, rather than relying solely on testing the end-product.

Eggs are a special case. In 1988, Edwina Currie, then UK Health Minister, declared that 'Most of the egg production in this country, sadly, is now affected with salmonella.' The reaction was ferocious and she was forced to resign. This was, at least in part, due to a misinterpretation of her actual remarks – she did *not* say that the majority of UK eggs were infected, which was not the case; one study estimated that only about one in a thousand eggs contained *Salmonella*. Although that is a low proportion, it has to be viewed in the context of a consumption of some 30 million eggs per day in the UK. What she should have said was that most of the places where eggs are produced are affected by *Salmonella*.

Let's look at the background to this story. When a flock is infected with *Salmonella* (in this case, specifically *S.* Enteritidis), in a few cases the bacteria are thought to infect the egg in the ovaries or oviduct, before the shell is laid down. In many cooking processes, such as soft boiling or light frying, the temperature within the yolk is not high enough to kill the bacteria. After the egg is eaten, the fat content shields the bacteria from the stomach acidity and, as with the chocolate bars, this means that a small number of bacteria can make someone ill. On a domestic scale, when an individual eats an egg for breakfast, a one in a thousand chance of becoming ill is a small risk. But on a catering scale, where large number of eggs may be used to produce a dish – say, mayonnaise – then the whole product becomes infected and many cases can result. The problem has not gone away completely, but it has been reduced in scale very extensively, largely due to vaccination of flocks and to the use of pasteurized egg in catering.

For many years, *Salmonella* was thought to be much the most common bacterium causing gastroenteritis in countries such as the UK. Subsequently, following the introduction of additional ways of growing bacteria from faecal specimens, it has been found that *Campylobacter* is at least as common, if not more so. The history of the discovery of *Campylobacter* has been covered in Chapter 3. Other bacteria that cause food poisoning are much less common. Staphylococcal food poisoning is due to a preformed toxin in the food. The classic example is a jug of custard left unrefrigerated overnight. If this contains a few *Staph. aureus*, they will grow in the custard and produce the toxin. The toxin is heat stable, so simply boiling the custard will not destroy the toxin, although it will kill the bacteria. When the custard is eaten, the toxin acts on the vomiting reflex mechanism in the stomach, causing violent vomiting.

Bacterial gastroenteritis is commonly caused by toxins affecting the secretion and uptake of fluid from the gut, often leading to watery diarrhoea (similar to, but not as severe as, cholera – see Chapter 3). Dysentery is rather different, in that the microbes invade the gut wall and set up inflammation. This causes severe abdominal pain and blood in the faeces. The main bacterial causes are *Shigella* species, the most severe of which are typically confined to tropical countries. However a milder form, due to *Shigella sonnei*, is not uncommon in countries such as the UK. As the infective dose is typically low (10–100 bacteria), the pathogens are readily spread from one person to another (as for viral gastroenteritis – see below), and can therefore cause problems in institutions such as schools and nurseries.

Bacteria that produce heat-resistant spores (mainly *Bacillus* and *Clostridium* species) can cause food poisoning as the spores will survive normal cooking, then grow as the food cools down. One bacterium from this group, *Cl. botulinum*, deserves a special mention because it causes a much more serious disease, botulism. Although originally described as arising from eating contaminated fermented sausages, this is most commonly a disease associated with canned foods. These are normally sterilized after canning but, if the temperatures inside the can are not high enough to kill the spores, cases of botulism arise. This is extremely serious, not only for the unfortunate patients but also for the manufacturers. In 1978, there was an outbreak of botulism in the UK arising from contaminated cans of salmon. Although there were only four confirmed cases (with two deaths), there was a dramatic decline in the sale of canned salmon (and other canned fish), with major economic consequences for the manufacturers. This problem is therefore taken very seriously.

Unlike the other diseases mentioned so far, botulism does not affect the gut. The disease is caused by a toxin which enters the nervous system and attacks the synapses – the connections between nerve cells where the impulses are transmitted from one cell to the next. Botulism specifically affects the transmission of impulses to the muscles, causing what is known as flaccid paralysis; the muscles are unable to contract. As this affects the respiratory muscles, the patient may die through being unable to breathe, so artificial ventilation is needed to keep the lungs working. Botulinum toxin has, in recent years, become familiar in an unexpected situation – the injection of tiny amounts of the toxin into the skin causes relaxation of the muscles locally, and hence the removal of facial wrinkles. As with the disease itself, the effect wears off after a while. The nerve cells themselves are not killed; only the synapses are affected. Eventually, the cells will make new connections to bypass the inhibited synapses, and function is restored after several months.

As a digression at this point, we can consider tetanus, which is caused by the related bacterium *Clostridium tetani*. This is not a food-borne disease, but it happens typically when tiny numbers of *Cl. tetani* get into a wound in the skin, such as may occur when pruning roses. Any wound, even an insignificant one, that is contaminated with soil (sports injuries, car accidents, etc.) can result in tetanus unless vaccination against it is up to date.

Tetanus is due to the action of a toxin that is very similar to the botulism toxin, but the symptoms are the opposite – the muscles contract uncontrollably, causing spasms. The effect on the facial muscles is especially characteristic, producing a 'sardonic grin', and the difficulty in moving the jaw leads to the common name 'lockjaw'. However, all muscles can be affected. The tetanus toxin also works by preventing transmission of nerve impulses at the synapses but, in contrast to the botulism toxin, it specifically affects *inhibitory* synapses. Since our muscles are controlled by a balance between stimulatory and inhibitory impulses, removing the latter means we have no way of preventing contraction of our muscles.

Returning to the subject of food poisoning, bacterial food poisoning tends to be more common in the summer, when it is warm enough for the bacteria to grow rapidly in the food, especially if left lying around for a while. Think of barbecues, cold chicken salads and so on. Next time you are at a wedding reception where the food is laid out ready while all the interminable speeches are being made and photographs taken, think of the bacteria multiplying within the food.

But not all gastroenteritis is like this. Some forms occur at all times of the year at similar frequencies, or even more often in winter. One form is known as 'winter vomiting disease'. These forms are typically due to viruses rather than bacteria, and the origins of infection are different. Although it may start with contaminated food, the main source of infection is from person to person. Whereas we usually have to ingest large numbers of *Salmonella* to become ill, with these viruses, much smaller numbers are sufficient. We can thus pick it up from touching contaminated door handles and other objects. The viruses survive better in the colder moister conditions in winter, and hence cause more problems at that time of year. Viral gastroenteritis is especially prominent amongst children. Each year, some three million children worldwide die from gastroenteritis caused by viruses such as rotavirus.

Another virus that receives much publicity is the Norwalk virus (now called norovirus), after Norwalk, Ohio, USA, where it was first characterized following an outbreak in a school in 1969. This virus crops up particularly on cruise ships. In February 2012, 364 passengers (out of 3,103) on the cruise ship *Crown Princess* fell ill with vomiting and diarrhoea. When the ship returned to Fort Lauderdale, Florida, it was subjected to sanitization and disinfection. Passengers then embarked for the next cruise, but unfortunately that had the same problem, with 288 of 3,078 passengers becoming ill. That cruise was terminated two days early to allow for extended sanitization.

This is just one example out of many, and the problem is by no means confined to cruise ships – it is just much more likely to show up with a group of people collected together for a week or two. Although the virus may start with food, it can easily be passed from one person to another because the infective dose is low. Once an outbreak has started, it is very difficult to stop it. All sorts of objects become contaminated – door handles, especially. Therefore, the only real solution is thorough disinfection and, as the example above shows, it does have to be *extremely* thorough.

Typhoid fever is rather different from the diseases we have looked at so far, in that the bacteria are invasive and get into the blood from the gut, causing fever and a rash, rather than symptoms associated with the gut. The bacteria subsequently return to the intestines, in large numbers, from the gall bladder, and hence can be passed on to other people. Before the advent of antibiotics, 10–20 per cent of people with typhoid fever died from the disease.

At the beginning of the 20th century, there was an outbreak of typhoid fever in a family in New York City. The cook, Mary Mallon, disappeared, and she was eventually tracked down by following a trail of further cases of the disease in other families for which she had worked. She was forcibly detained by the New York City Health Department, but was released after giving an undertaking not to work as a cook again. However, in 1915, she was again working as a cook, at a hospital in New York, where there was an outbreak affecting 25 people, two of whom died. She was eventually detained in a hospital until her death, in 1938.

The case of 'Typhoid Mary' is only the most famous example of a characteristic problem with typhoid fever. The bacterium that causes it, *Salmonella typhi*, can persist in the body (especially in the gall bladder) for years, without any symptoms, so the carrier is perfectly healthy – except that they release bacteria into their faeces, from where they can contaminate food.

Note that typhoid fever is totally different from typhus, with which it is often confused. Typhus is a louse-borne disease caused by an unusual bacterium known as *Rickettsia*.

In addition to bacteria and viruses, protozoa are also involved in food and water-borne diseases, such as amoebic dysentery. Although the most severe forms are found in tropical countries, protozoa causing dysentery are not absent from Europe. *Giardia* is the main example, usually from water rather than from food, but it can occur on salad vegetables and fruit such as strawberries. It is usually associated with washing the produce in contaminated water. The cysts of *Giardia* are resistant to chlorination, but are readily killed by cooking.

Another water-borne protozoan that causes a lot of concern is *Cryptosporidium*. This is common in animals such as cattle and sheep, and it can cause major outbreaks if it gets into the water supply. In 1993, an outbreak in Milwaukee, Wisconsin, USA caused over 400,000 cases. In most people, the disease is self-limiting (although it can cause diarrhoea lasting up to 20 days) but, in individuals with AIDS, it can be life-threatening.

Finally, we need to come back to bacteria, and *E. coli* in particular. This is a normal gut inhabitant, and most of the time it causes no problems. It has been the mainstay of bacterial research from the outset, especially in the fields of genetics and biochemistry, and it is widely used in student laboratory classes. However, some strains can be pathogenic, and these cause a wide variety of diseases. Some strains produce a toxin related to the cholera toxin, and cause a cholera-like disease (albeit milder). Other strains resemble *Shigella*, and cause a type of dysentery. However, the strains that grab the headlines are those with the cumbersome title of enterohaemmorhagic *E. coli* (EHEC), more familiarly known as *E. coli* O157:H7 (which refers to the types of antigen it possesses – although other types also share the dangerous properties of this strain). The main life-threatening condition caused by this strain is called haemolytic uraemic syndrome (HUS), which is marked by acute renal failure, plus other symptoms, and has a mortality rate of 5–10 per cent.

A very severe outbreak of EHEC occurred in Europe (mainly in Germany) in May and June 2011. The first recorded case was on 1 May, but it was three weeks before the cause was identified as an unusual EHEC strain (specifically known as O104: H4). By the end of the outbreak, there had been nearly 4,000 cases in Germany, with 54 deaths.

The initial epidemiological evidence suggested (incorrectly) that Spanish cucumbers were the source, which led to substantial economic losses to the Spanish farmers (and others elsewhere), as enormous quantities of produce were dumped as unsalable. Ultimately, the source was traced to contaminated bean seeds and the practice of germinating these seeds in a warm, moist environment (perfect for bacterial growth and survival); the sprouted seeds were then sold as ready to eat. A separate, smaller, outbreak in France (15 cases) in June 2011 was traced to the same original source (imported fenugreek seeds), but a different sprout farm. The outbreak in Germany may have been exacerbated by re-contamination from two workers at the German sprout farm who were identified as excreting *E. coli* O104:H4.

We have now seen a number of examples of the roles of microbes in health and disease. The next chapter represents a change of scene – looking at their roles in the environment.

6
Microbes and the Environment

Microbes are everywhere in the environment. However, this rather glib statement needs some qualification and explanation. Conventionally, we can start by thinking of 'the environment' as consisting of soil and water – fresh water (streams, ponds and lakes) and salt water. Even the air above us contains large numbers of microbes, especially in drops of water within clouds, where they may play a role in fixing atmospheric nitrogen and the transformation of polluting chemicals. There are also a lot of larger organisms (including ourselves), and these provide quite distinct environments for microbial growth and survival, as bacteria and other microbes can live in and on these larger organisms. So, we have to add in plants and animals as part of the natural environment for microorganisms.

The environment is not just a passive vehicle for microbial life; the microbes interact with one another, and with other forms of life, in a complex manner. Furthermore, the environment at a microscopic level can be very different over a distance of just a millimetre or so. For example, the human mouth would appear to be an aerobic environment, i.e. one that is freely exposed to the air – yet, as we saw in Chapter 2, hiding away in crevices between the teeth are bacteria so sensitive to oxygen that they die quickly if exposed to air. These localized microenvironments as referred to as 'niches'.

There are several themes that will echo throughout this chapter. One is the scale of microbial activity and diversity in the environment. Although microbes are very small, their numbers are so large that they represent perhaps 50 per cent of the total biomass of the Earth, and they contribute as much to the total photosynthetic ability of the planet as all land plants combined. Their diversity has only recently been fully appreciated. We now know that conventional methods that involve trying to grow microbes in the laboratory recover only a very small proportion of the species

Understanding Microbes: An Introduction to a Small World, First Edition. Jeremy W. Dale.
© 2013 John Wiley & Sons, Ltd. Published 2013 by John Wiley & Sons, Ltd.

present. DNA studies have shown the presence of many more, previously unknown, organisms (see Chapter 2, and later in this chapter).

The second theme is that studies of environmental microbes have caused a radical shift in our concept of the conditions under which 'life' is possible. Microbes have been found in the most extraordinary places – extremes of pH, salinity, and temperature. I will return to these issues later in the chapter. In Chapter 10, I will consider some of the implications of this for our search for life elsewhere in the universe.

A further recurring issue is what these microbes use for food, and what do they do with this material? Here, I want to recapitulate and expand on the topics covered in Chapter 1. The metabolic activity of microbes – how they get their food and what they do with it – is central to the impact of microbes on the environment, and vice versa. All forms of life require a number of factors, but for present purposes the key ones are sources of carbon, nitrogen and energy. Microbes play a vital role in the cycling of carbon and nitrogen, without which life as we know it would be impossible. A bacterium like *E. coli*, growing in the lab, can use simple sugars as a source of both carbon and energy, and ammonium salts as a nitrogen source.

In the wider environment, things are often not that simple. Some bacteria – notably the cyanobacteria – and most algae are photosynthetic, so they can use sunlight as a source of energy. Since they do not need to use sugars or other organic compounds as an energy source, they can take carbon dioxide from the air as their carbon source and can construct sugars and other organic compounds from that – so they fix carbon in much the same way as plants do. This plays a vital role in the carbon cycle, converting carbon dioxide (CO_2) into biomass (see Figure 6.1a). As already mentioned, although they are very small organisms, and often unseen, their numbers in lakes and oceans are vast and they are as important as plants in taking CO_2 out of the air (and hence helping to counteract the effect of our release of greenhouse gases). At the same time, other bacteria have the opposite effect. They break down organic matter, releasing CO_2 into the air if they are growing in the presence of oxygen. In the absence of oxygen, such as in sediment at the bottom of ponds or lakes, they may produce methane instead, which is an even more potent greenhouse gas than CO_2.

The second key requirement is the need for nitrogen (see Figure 6.1b). This is plentiful in the air, where it makes up about 80 per cent of the air we breathe, but we cannot use it in that form, and nor can most other organisms. However, some bacteria can. You may be familiar with the nitrogen-fixing ability of leguminous plants (peas, beans, clover and so on). This is due to a bacterium known as *Rhizobium*, which occurs in nodules on the roots of these plants. Rhizobia produce an enzyme (nitrogenase) which converts nitrogen to ammonia, which the plant can use. The bacteria acquire the necessary energy, as well as a carbon source, from the plant, in the form of organic compounds. Excess ammonia is released into the soil, where other microbes can oxidize it first to nitrite and, from there, to nitrate (nitrification).

(a)

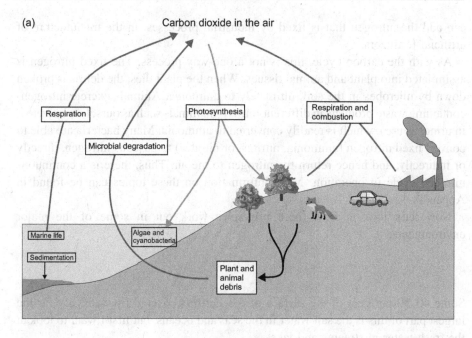

(b)

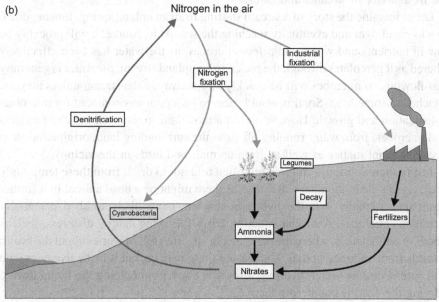

Figure 6.1   (a) Carbon cycle; (b) nitrogen cycle.

Rhizobia are not the only bacteria that can fix nitrogen. Free-living organisms, notably cyanobacteria, are also important in this respect, especially in the seas. Nitrogen-fixing bacteria are by far the most important natural source of fixed nitrogen. A small amount comes from lightning discharges as well, and, to this, we can add the nitrogen that is fixed by industrial processes, in the manufacture of artificial fertilizers.

As with the carbon cycle, this is not a one-way process. The fixed nitrogen is assimilated into plant and animal tissues. When the plant dies, the debris is broken down by microbes in the soil, ultimately to ammonia. Animals excrete nitrogen-containing waste products in different forms – mammals such as ourselves excrete it in urine, as urea, which is readily converted to ammonia. Many bacteria are able to convert fixed nitrogen (ammonia, nitrites, or nitrates) to gaseous nitrogen, directly or indirectly, and hence return the nitrogen to the air. Thus, there is a continuous nitrogen cycle in operation. More information on these topics can be found in Appendix 1.

Now let's look at how these principles work out in some of the major environments.

6.1   Water

Some 40–50 per cent of the surface of the Earth is covered by water. By far the largest part of this is the salt water in the seas and oceans, but first I want to look at the fresh water in streams and lakes.

Let us imagine the story of a stream starting from an upland spring, flowing down to a lowland river and eventually reaching the sea. At its source, it will probably be low in nutrients and will contain few microbes, as the water has been effectively filtered as it percolates through the rocks. In an upland stream, the water is generally fast-flowing, so microbes will be quickly swept away in the stream unless they are attached to something. So, this would seem to be a poor environment for microbial colonization and growth. However, even a mountain stream will acquire nutrients and microbes, from water running off from the surrounding land, originating from decaying plant matter as well as from animals and birds in the vicinity.

Just to digress briefly, children are often told not to drink from these temptingly cool, crystal-clear, mountain streams, 'as there might be a dead animal in it further upstream'. Actually it is the live ones that are more likely to shed potentially hazardous microbes. A similar situation arises following natural disasters, such as floods or earthquakes, where the news media are often full of stories about the health hazards from unburied bodies. Unpleasant these may be, but it is the live ones who will spread disease – and diverting resources away from helping the living toward burying the dead is counterproductive.

Returning to our stream, although many of the microbes will simply be swept away by the flow of water, some will be able to attach to rocks and plant surfaces in the water, forming a layer of organisms encased in organic matter – this is known as a biofilm (see also Chapter 8). Within these communities, there are likely to be photosynthetic bacteria such as cyanobacteria, as well as algae, which are able to supplement the low levels of nutrient in the water by using sunlight to produce the energy they need. There may be other pigmented bacteria which are not photo-synthetic; the pigments provide a degree of protection against ultraviolet radiation which can penetrate the shallow water. As these communities develop, other organisms move in, taking advantage of the nutrients provided by the original colonizers, including protozoa that feed by grazing on the biofilms.

As the stream flows downhill and reaches the lowlands, joining up with other streams, the flow slows down, and more run-off joins from neighbouring agricultural land – together with, ultimately, sewage outfalls. This leads to a build-up of nutrients, especially nitrogen-containing material such as nitrates (from fertilizer) and other compounds from animal waste. This can encourage the development, especially in lakes and ponds, of surface blooms of photosynthetic organisms (algae and cyanobacteria). This process, known as eutrophication, is extremely damaging, especially as it can result in the water becoming de-oxygenated, ultimately causing the death of fish. At the same time, the slower flow allows sediment to build up. The resulting turbidity makes the water less favourable for photosynthetic microbes, although they will still occur at the margins and on the surface. On the other hand, other microbes can multiply, taking advantage of the nutrients available.

Figure 6.2 *Dalton collecting marsh fire gas*, Ford Madox Brown (1821–1893). Detail from a mural in Manchester Town Hall. John Dalton (1766–1844) stirs the bottom of a pond while his assistant collects the gas in an inverted jar. (Reproduced by permission of Manchester City Council)

The mud at the bottom of a lake or pond is a very different environment from the water above it. Not only is it far from the surface, so that little air penetrates, but also the microbial decay of vegetation and other matter that sinks to the bottom will use up whatever oxygen there is. This is essentially an anaerobic environment, and the organisms that thrive there are adapted to life without air. One consequence is that their metabolism produces incompletely oxidized products, notably methane and hydrogen sulphide (H_2S) (it is the latter, and similar compounds, that are responsible for the smell – methane itself is odourless). The results can readily be seen when the bottom of a pond is stirred up – bubbles of gas are released.

One of these anaerobic bacteria is especially significant, namely *Clostridium botulinum*, which causes botulism. People do not (usually) eat mud, so it is not a human problem. However, it can be a problem for birds, especially ducks and waders that feed in the mud at the bottom or edges of lakes or ponds, and it can cause devastating outbreaks among bird populations.

Eventually, the stream or river, reaches the sea. This is now a very different story – or rather a collection of different stories, as it encompasses a number of very different habitats. Two factors unite these: the presence of salt and a low level of nutrients. Of course, this does not apply where the rivers join the sea, as the water is less salty and also still contains the nutrients and microbes supplied by the river. Sewage outfalls also supply nutrients and microbes. In the open sea, however, these nutrients get used up by organisms both large and small, and these organisms eventually die and sink to the bottom. The depth of the sea means that, in contrast to lakes and shallow waters, there is little recirculation of nutrients from those deposits. This is an important factor when considering the amount of carbon dioxide in the air, and hence the question of climate change. Rather than the carbon being recycled, as in the above discussion of the carbon cycle, it is deposited at the bottom of the ocean. This is estimated to account for as much as 9 billion tonnes of CO_2 per year, or about a third of the total produced.

The consequence of the low nutrient level in the seas is that much of the primary activity is due to photosynthetic organisms, including cyanobacteria and algae, that get energy from sunlight and fix carbon and nitrogen from the air. However, they do need other nutrients, such as iron and phosphorus, and the availability of these is an important limiting factor. The main supply of minerals to the oceans is in dust from the air. For example, dust storms arising from deserts such as the Sahara provide much of the iron that feeds the microbes in the oceans. There has been much speculation, and controversy, about the possibility of adding iron to areas of ocean in order to stimulate microbial activity, and thereby fix more carbon dioxide from the air, as a way of reducing the greenhouse effect. However, limited trials have not been encouraging.

As these organisms need light and air, their main activity is at the sea surface. Therefore, in the shallower zones around the edge where light can penetrate, photosynthetic organisms can thrive and sometimes multiply uncontrollably. This

Figure 6.3 An algal bloom in the waters of La Jolla, California, USA. (Photo credit: Kai Schumann/National Ocean Service, US NOAA)

uncontrolled growth can cause 'blooms', which may happen regularly or sporadically, and are especially associated with high levels of nutrients due to run-off from the land or sewage outfalls.

Some blooms are quite colourful – white blooms, for example, arise from algae known as coccolithophores, which have a calcium-containing skeleton, while the so-called 'red tides' are due to dinoflagellate microbes. Although these might seem to be rather pretty, they are damaging because they exclude light from the water. Also, the deposition of large amounts of rotting algae on the shoreline is extremely unpleasant – and sometimes dangerous, as the decaying matter can release toxic gases. For example, in July, 2011, a number of wild boars were found dead on the muddy banks of the Gouessant estuary in Brittany, surrounded by piles of rotting algae. Beaches were closed as bulldozers were brought in to clear the stuff away. And this was not the first such incident; two years earlier, in a neighbouring part of Brittany, a 27 year old man was dragged unconscious from a metre-deep patch of rotting algae after his horse had collapsed and died.

Some blooms can be extremely damaging to wildlife in other ways, and even to humans, due to powerful toxins produced by the microbes. One of these microbes, a dinoflagellate called *Pfiesteria*, causes blooms typically in partly enclosed bays and estuaries, where sewage run-off can lead to high nutrient levels, When fish come into contact with such a bloom, they can die within a minute. This organism can also be extremely dangerous for scientists working with it in the lab, causing a variety of severe nervous system damage ranging from delirium to memory loss. Symptoms can persist for years (if the victims survive). Furthermore, shellfish that feed on these and other algae can accumulate toxins which result in problems when people eat the contaminated shellfish.

These blooms typically disappear after a while. This may be due partly to the uncontrolled expansion of numbers using up the available nutrients, but it can also be due to the activity of viruses or other parasites. If we think of these as a form of predator, and the organisms in the bloom as their prey, we can see parallels with many situations. When predators are few or absent, the organism multiplies; this provides more food for the predators, so their numbers increase until there are enough of them to check further growth of the prey, or even to eliminate it – with the consequent collapse of the predator population as well. Large predators tend to have longer generation times than their prey, so predator numbers will respond more slowly (unless they are attracted from surrounding areas). But smaller 'predators' such as viruses can multiply more rapidly than their host, and this will cause the sudden collapse of a bloom.

Before the use of DNA analysis, the only microbes we knew about were those that could be grown in the laboratory. Now we know that this is just the tip of the iceberg – and, in fact, very much less than that. In Chapter 2, I said that less than 20 per cent of the bacteria in the intestines had previously been grown in the lab. In seawater samples, the situation is more extreme, with less than one per cent of the bacteria being previously known.

The evidence for this is obtained using methods that are described more fully in the next chapter. Briefly, samples of seawater are taken and all the DNA present in whatever organisms the water contains are extracted, without attempting to separate the organisms in any way. The technique known as polymerase chain reaction (PCR) is then used to make a lot of copies of a specific gene from all the bacteria present – a process called amplification. Usually, the chosen gene is that coding for the ribosomal RNA (rRNA), as this will be present, and very similar (but not identical), in all bacteria. These amplified fragments can then be cloned, using *E. coli* as the host. As described in the next chapter, this results in thousands of *E. coli* clones, each carrying a different bit of DNA from the seawater, since each clone will carry a single piece of DNA. Between them, these clones represent all the bacteria present in the sample. From each of these clones, the cloned fragment can be sequenced and compared to the rRNA sequences of known organisms.

Advances in DNA sequencing technology have now made a more sophisticated approach possible. All the DNA in the sample, not just that coding for the rRNA, can be broken into relatively short pieces, and each fragment can be sequenced. Piecing together the sequences of these random fragments means that the genome sequences of many of these organisms can be determined, at least in part. Furthermore, from that data, it is possible to predict their properties, such as whether they are photosynthetic. In some cases, it was predicted that a specific microbe required a key chemical for growth, due to the absence of the genes needed to make it. That enabled that microbe to be isolated and grown in the lab. A pioneering study in 2004, using samples from the Sargasso sea, obtained partial genome sequences from 1,800 different species, many of them previously totally unknown. These methods, known as *metagenomics*, have applications in many fields.

Another surprise is the number of viruses that are present in the sea. These are not the viruses that cause human disease (although those can be found in coastal waters

around sewage outfalls), but viruses that infect bacteria (which are called bacteriophages, or phages for short) or infect other microbes, such as algae. These viruses
can be present in vast numbers, up to 10^7 per ml in the top 50 metres. Factoring in the
size of the oceans (which cover some 40 per cent of the Earth's surface) and making
similar estimates for the number of phages in soil and other environments, it has
been calculated that there may be about 10^{31} bacteriophages in the world in total,
making them the most numerous form of 'life'.

Although these viruses are very small, their sheer numbers make the total mass
surprisingly large – estimated at 10^9 tonnes, or a thousand million tonnes.
However, that does need to be set against the total global biomass of over a
thousand *billion* tonnes (of which more than half is contributed by bacteria). Many
of these viruses do no damage to their hosts but live in a more or less stable
relationship with them (see the description of lysogeny in Chapter 1). Even so,
they may play a significant role; some of the viruses that infect cyanobacteria, for
example, carry genes for important elements of the photosynthetic machinery.
On the other hand, some viruses do cause damage to their hosts, which is most
dramatically shown by the sudden collapse of blooms due to algae or cyanobacteria, as described earlier.

Below the surface of the sea, about 75 m down, the temperature drops rapidly
over a short distance – this is known as the thermocline. Below this point, it is not
only cold but also permanently dark, so photosynthesis is not possible. The
relatively few organisms present here have to use nutrients obtained from particles
that are gradually sinking from the surface layers of the sea.

In the deep layers, things get very interesting. Although we would regard the
environment as very inhospitable. Not only is it cold (in general), dark and nutrient-
poor, but also the pressures in the deepest trenches are immense, reaching over a
thousand times the pressure that we experience. The most interesting features are
the thermal vents – known as 'black smokers' – in the ocean floor. These can be
thought of as similar to partially dormant underwater volcanoes, where very hot
material (at temperatures up to 400 °C) emerges from the depths of the Earth and
mixes with the seawater. This heats the seawater to temperatures higher than the
'normal' boiling point of water (due to the very high pressures at those depths, the
water does not boil). Surprisingly, specialized microbes (bacteria or Archaea)
are able to grow in these regions. These microbes are known as 'thermophiles'
(from the Greek, meaning 'heat-loving').

At the high temperatures in and around the vents, seawater reacts with the rocks,
producing gases such as hydrogen, hydrogen sulphide (H_2S) and methane. The
microbes that live around the vents can obtain their energy by oxidizing these gases
to water, sulphate or carbon dioxide, respectively. Many of the larger organisms that
live in these places, such as giant tubeworms, mussels and shrimps, exploit these
microbes through a symbiotic relationship. They contain a permanent population of
bacteria that are able to use these gases to get the energy that they need and their
hosts, in turn, use the bacteria to provide their food.

Figure 6.4   A deep ocean vent. A bed of tube worms cover the base of the black smoker. Environmental testing devices are also in the picture. (*Source:* Vents Program, Pacific Marine Environmental Laboratory, US NOAA URL: http://www.pmel.noaa.gov/vents/gallery/smoker-images.html)

Part of the reason for the ability of these thermophiles to survive and grow at high temperatures is that they have evolved enzymes that can tolerate such conditions without being inactivated. This has real significance for practical applications in biotechnology. Many processes that are commercially important, or potentially so, use enzymes to carry out specific reactions. With conventional sources of these proteins, the temperature at which the reaction occurs is restricted by the inability of the enzyme to survive at higher temperatures. If an enzyme from a thermophilic microbe is used instead, then the reaction can be carried out at high temperatures, which is much more efficient. I will come back to this again in Chapter 9.

Another class of specialized microbes is found in the depths of the sea. The salt concentration in the sea is not uniform; where there are deep pockets in the sea, the salt concentration may be much higher. This makes the water more dense, so it stays at the bottom and does not mix with the layers above it. Such pockets exist, for example, in the Red Sea, the Gulf of Mexico and the Mediterranean. Microbes that are found in these regions have to be able to tolerate, and grow in, water with high levels of salt. They are known as halophilic (salt-loving) microbes.

We'll look further at thermophiles and halophiles, and others that live in extreme environments, later in this chapter. But before doing that, there are other environments to look at, starting with soil.

6.2 Soil

Soil is extremely varied and complex. At one extreme, we have sand, which is often very dry and poor in nutrients and thus has a limited number of microbes. At the other extreme, we have clay, which is also rather inhospitable. Even if we limit ourselves to a reasonably good garden soil, and just consider the top layer (say, down to a spade's depth), conditions will vary considerably, especially in temperature and wetness – to say nothing of the complexity of the microenvironments within that spadeful of earth. As might be expected, that diversity of environments is reflected by very different microbes, so this discussion will necessarily be somewhat generalized and superficial. Let's confine ourselves to looking at a few of what I consider are the most interesting aspects of soil.

The first thing that will probably strike the human senses is the smell – especially after a shower of rain on warm soil. The characteristic smell of damp earth is due to one of the most important groups of microbes in soil, *Streptomyces*. These belong to a group of bacteria known as the actinomycetes, and they grow in long filaments called *hyphae* rather than as individual cells. The network of branching and interconnecting hyphae is called a mycelium. This method of growth is similar to that of fungi (although the hyphae of *Streptomyces* are much finer), but these are not fungi but really are bacteria. As with the fungi, the network of hyphae enables *Streptomyces* to access nutrients that may only be present in small pockets within the soil.

One reason for being especially interested in the actinomycetes is that they are the major source of naturally-occurring antibiotics, as well as other compounds that have therapeutic uses, including anti-cancer drugs and immunosuppressants. *Streptomyces* and the other actinomycetes produce an amazing array of chemicals – known as secondary metabolites – that seem to have little or nothing to do with their normal processes of growth and replication. Exactly why they do this is something of a mystery. The antibiotics may play a role in reducing competition from other microbes, but there is little evidence that this gives them a really important selective advantage. Another possibility is that these are really signalling compounds, enabling the organism to coordinate its activities, and their therapeutic usefulness is merely an accidental by-product.

The most familiar mycelial microbes within the soil are the fungi, although it is not the network of filaments under the ground that are most recognizable but the mushrooms and toadstools that are the fruiting bodies produced by that mycelium. When conditions are right, a sexual process occurs within the mycelium, resulting in the cells differentiating into the complex structures that we see above the ground. This is a mechanism for both reproduction and spread. Large numbers of spores are produced in the cap, from where they are dispersed to start new hyphae elsewhere.

Life in the soil is not a bed of roses. Bacteria are at the bottom of a food chain. All their hard work in obtaining nutrients from the soil and air, including fixing carbon and nitrogen, is taken advantage of by predators, the most significant of these being the protozoa. Many species of protozoa live in soil, moving around in the films of water that surround soil particles. Some can use a variety of nutrients, but generally their main source of food is bacteria, which they engulf and digest.

However, not all bacteria are killed by protozoa. Some bacteria are able to survive ingestion, and even multiply within the protozoa. They have evolved mechanisms that enable them to resist the killing action of the protozoa. A notable example is *Legionella*, which causes Legionnaire's disease (see Chapter 3). I will return to the subject of intracellular bacteria and biofilms in Chapter 8, as well as other interesting organisms in soil and water, such as myxobacteria and slime moulds, which show aspects of multicellular behaviour.

Beyond the protozoa of course, there is a complete food chain – multicellular animals such as nematodes eat the protozoa (and also bacteria), and are in turn eaten by earthworms, and so on. But to go into that would extend beyond the world of microbiology.

6.3   Plants

One aspect of the interaction of microbes with plants, the contribution of bacteria such as *Rhizobium* to nitrogen fixation, has already been mentioned earlier in this chapter. But that is far from the end of the story. In order for a plant to reach the water and nutrients it needs, it has to send its roots out far and wide; and so that it can penetrate into and between the crevices in the soil, there are hairs attached to the roots. However, on a microscopic scale, even those hairs are rather clumsy. Fungi can do it much better. Many plants rely, therefore, at least partly, on an association with fungi, known as a mycorrhizal association. The mycorrhizal fungus not only spreads outwards, gathering in nutrients, but also penetrates the structure of the plant roots, either by pushing their way in between the cells in the root or by actually entering the plant cells.

Mycorrhizae are more common than is generally appreciated. A field guide to the identification of fungi is likely to indicate that certain species are commonly found in the neighbourhood of specific trees – beech, oak, etc. What we are trying to identify turns out to be the fruiting body of a fungus that has a mycorrhizal association with that type of tree.

Mycorrhizal fungi can be very important for plant health, especially in nutrient-poor soils. They are particularly useful in releasing phosphorus and nitrogen from sources that would otherwise be inaccessible to the plant, as they are able to degrade complex materials that the plant cannot. Also, by searching out water efficiently, they can make a plant more drought-tolerant.

Ericaceous plants – including not only heathers but also commercially important plants such as blueberries – provide one example of the importance of mycorrhizae. These plants commonly grow, or are grown, in soil that is relatively nutrient-poor, although rich in peat or humus. If they are introduced into a new site where ericaceous plants have not previously been grown, they may fail – or at least take a long time to get established – because they lack the appropriate mycorrhizae that would help them to acquire food and water. Artificial introduction of mycorrhizal fungi can be beneficial in such circumstances.

However, microbes are not all beneficial to plants. There are very many diseases caused by microbes of all kinds (ignoring those that are caused by insects and other pests). As with human infections, microbes that infect plants have to get around a set of defence mechanisms, of which the most important is the relatively tough outer layer of plants and their cells. So anything that damages the plant, whether it is a pruning wound, other mechanical damage or penetration by a sap-sucking insect, will make the plant more susceptible to disease.

In Chapter 2, I described two aspects of the immune system in mammals, broadly separated into adaptive immunity (antibodies and cell-mediated immunity) and innate immunity. Although plants seem to lack adaptive immunity, they do have an innate immune system which resembles some aspects of that seen in mammals, especially in recognizing specific chemical structures, known as PAMPs (pathogen-associated molecular patterns), which are common to many invading microbes. Many plant pathogens can avoid this protective mechanism by producing proteins that suppress the plant's innate immunity. The contest can then continue by the plant switching on a further defence mechanism, which may include the production of antimicrobial chemicals or death of the infected cells. Let us look at just a few examples of plant diseases.

From time to time, diseases occur which are more than a personal tragedy for those concerned, but leave a long-lasting scar on human history. I have already looked at some examples of human diseases in this respect. Here is another example: in this case not a human disease, but a disease of plants and, specifically, the potato.

In the early part of the 19th century, up to a half of the population of Ireland was almost entirely dependent on the potato for their food. When the potato crop failed, as it did intermittently in some areas, there was starvation. But when a new pathogen, *Phytophthera infestans*, arrived in Ireland, probably in 1844, the scale of the problem was much larger than any that had been seen before. In 1845, up to 50 per cent of the crop was lost, and in 1846 the loss was about 75 per cent. Severe crop losses continued in subsequent years.

It is estimated that, over the period between 1846 and 1851, about one million people died, either of starvation or of diseases linked to starvation, and a further million emigrated. Since the population of Ireland (in 1841) was about eight million, this represents a drop in population of about 25 per cent over five years – and the proportion was higher than that in some parts of western Ireland. Although the epidemic eventually declined, as epidemics tend to do, poverty did not stop with the end of the famine and neither did emigration, so that, by the turn of the century, the population of Ireland was little more than half what it had been before the famine.

The organism that caused the infection, *Phytophthera infestans*, although commonly described as a fungus, is actually a member of the oomycetes, or water moulds. These were originally thought to be fungi, but are now known to be quite distinct and more closely related to the algae, although they have lost the ability to photosynthesize.

Although this organism was responsible for the loss of the potato crop, the causes of the famine run much deeper than that, and include the political and economic conditions in Ireland at the time and the response of the British government (which ruled Ireland then). Even at the height of the crisis, the large landowners who controlled the best agricultural land were exporting grain and cattle to England, and the British government was resistant to suggestions that this trade should be stopped so the food could be used to alleviate starvation in Ireland. Nor did they provide effective aid, and they even attempted to hinder others from doing so. The reasons for these actions are complex, and this is not the place to go into what is still a highly charged political issue, but it is undeniable that it provided extensive fuel to a growing Irish nationalism, with repercussions that came to a head some 70 years later and resounded throughout the 20th century.

Oomycetes also cause significant diseases of trees. One oomycete, *Phytophthera ramorum*, which is believed to have originated in rhododendrons, had its major impact in species of oak trees ('sudden oak death') in western USA (California and Oregon), where it caused the loss of several million trees. More recently, a variant of this pathogen has caused extensive loss of a wider range or trees; outbreaks in the UK in 2009–10 are estimated to have affected 1,900 hectares of larch plantations, with the loss of half a million trees.

For one of the best known diseases of trees, we can look at a real fungus, *Ophiostoma ulmi*, which is the cause of Dutch elm disease. This disease first appeared in Britain in the 1920s. While it caused a loss of a substantial number of elms at that time, the effect was less than had been originally feared, and by 1940 the epidemic had died down. However, in the late 1960s, a more aggressive fungus (*Ophiostoma novo-ulmi*) appeared which led to the death of most elms in southern Britain. The epidemic has been slower to spread in the north of the country, partly because of the climate and partly because of differences in the preponderant species of elm, but it is still progressing.

The disease is spread from one tree to another by the elm bark beetle. As a reaction to the fungus, the xylem channels become plugged, which prevents water and nutrients reaching the leaves. Although the roots are not directly affected, they are starved of the nutrients produced by the leaves, and they may therefore die. However, that does not always happen – the roots may survive and send up suckers which grow for a number of years before succumbing in turn. So, some young elm trees may still be seen in the hedgerows, but they are ultimately doomed.

There are many other fungi that cause a wide variety of plant diseases. Notable are the fungi (mainly of the genus *Puccinia*) that cause different forms of disease known descriptively as 'rust', because of the yellow-orange powdery pustules on the surface of the stem or leaves. This powder actually consists of the spores of the fungus, in which form it can disperse from one plant to another. When they land on a new plant, the spores germinate and penetrate the tissues of the plant. The rust fungi are obligate pathogens – they depend on their host for completion of their life cycle, and some have a complex life cycle involving alternate plant hosts. Although many plants are susceptible to one form of rust or another, this group of diseases is especially important with cereals. One of these diseases, stem rust of wheat, is worth considering further as an example.

Stem rust, caused by *Puccinia graminis,* has been a scourge for thousands of years and may cause the loss of 80 per cent of the crop. In the 1960s, work at the International Maize and Wheat Improvement Center (CIMMYT), based in Mexico City, lead to the introduction of a wheat strain that was resistant to stem rust (for which Norman Borlaug received the Nobel Prize in 1970). This strain was subsequently used throughout the world, and this seemed to have solved the problem – to the extent that further research on this disease was shut down. In 1998, however, scientists in Uganda noticed that these apparently resistant plants were being affected by a new strain of stem rust, known as Ug99; the pathogen had evolved a mechanism for overcoming the resistance gene in the wheat.

The problem arose because the resistance of the wheat was due to a single gene, known as *Sr31.* Although this conferred effective resistance, the presence of a single resistance mechanism made it possible for mutation of the pathogen to evolve a corresponding mechanism for overcoming it. Developing further new wheat strains in the same way would be likely to have the same outcome. An alternative is to breed varieties that have a number of different resistance genes, each of which only confers a small amount of resistance, but where the combination makes for effective resistance to the pathogen. Evolution of the pathogen to develop resistance to all of the battery of genes simultaneously is much less likely.

It is not only fungi that cause diseases in plants – bacteria and viruses are also important. One bacterial disease that is particularly interesting is crown gall, caused by *Agrobacterium tumefaciens.* This produces a tumour-like growth on the plant, which is due to the transfer of part of the DNA of a plasmid (known as a Ti, or tumour-inducing plasmid) from the bacterium into a plant cell where it integrates into the chromosome. This results in the unregulated expression of some plant hormones within the plant, causing the characteristic growths. The special interest of this arises from the fact that it provides a way of introducing new genes into the plant, by attaching them to the plasmid, and this is one of the techniques used for the genetic modification of plants. The Ti plasmid used for this has been manipulated to stop it causing disease.

Plant viruses are typically spread by insects such as aphids, rather than infecting plants directly. Vertical transmission (i.e. transmission to progeny rather than from one plant to another) also happens. This is an important consideration for plants that are propagated asexually (e.g. from cuttings, or from bulbs or tubers – such as

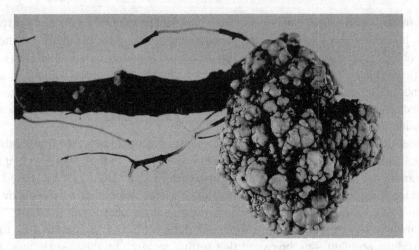

Figure 6.5   Crown gall on apple. (*Source:* Melodie Putnam, Oregon State University Plant Clinic, 2010)

potatoes – rather than from seeds). If the original plant was infected by a virus, then all the cuttings or tubers will carry the virus as well. So, if potatoes are being grown, it is important that the seed potatoes are derived from virus-free stock.

A viral infection of plants in general is often recognizable by yellow or black spots on the leaves, or stripes of different colour in the flowers. The latter is seen most notably in tulips, where variegated flower colours, which are desirable traits, are often due to a virus.

One plant virus that is particularly significant historically is the tobacco mosaic virus, which causes mottled patterns of the leaves of tobacco and other plants (including tomato and pepper), leading to significant economic losses. This was the first virus to be identified and characterized. In 1898, Martinus Beijerinck observed that the disease was caused by a 'contagious living fluid', and that whatever caused the disease was able to pass through a filter that was too fine for bacteria to pass through. He proposed this ability to pass through a fine filter as the defining characteristic of a virus. The viral particles were crystallized in 1935 by Wendell Stanley, and in 1936 they were shown by Norman Pirie and Frederick Bawden to consist of RNA with a protein coat. In 1955, Heinz Fraenkel-Conrat showed that the RNA itself was infectious.

6.4   Biodegradation

Biodegradation of plant debris in the soil is vital for the environment. Up to 70 per cent of plant material consists of cellulose, an insoluble polymer which can only be broken down by microbial activity. A variety of fungi and bacteria can do this, their relative importance depending on the soil type. Without this activity, the carbon cycle would come to a grinding halt, and our environment would fill up with unrotted plant material.

Woody plants contain even more resistant polymers, such as lignin (20–30 per cent of woody material). However some microbes can also digest lignin. These are primarily associated with rotting of fallen wood – although some, such as *Armillaria*, can also attack living plants and are, therefore, significant pathogens of trees and other woody plants. Although degradation of lignin is most commonly due to fungi, some bacteria (especially *Streptomyces*) are also capable of doing this. This activity has potential applications, for example in the generation of biofuels from woody material (see Chapter 9), Unfortunately, this sort of activity also has unwanted consequences, in the rotting of fence posts and structural timbers inside and outside the house, which I will come to shortly.

The biodegradative activity of microbes can be greatly enhanced by accumulating plant wastes, and food wastes, in a compost heap. Naturally occurring microbes, both fungi and bacteria, will start to break the material down, and their activity will cause the temperature to rise. Above about 40 °C, the initial organisms will be replaced by thermophiles (temperature-loving microbes that will continue to grow at high temperatures). These are mainly bacteria (including actinomycetes). Although fungi persist in a compost heap, they are mainly found around the edges where the temperature is lower.

In commercial composting, temperatures of 65 °C or more can be achieved. The high temperatures not only speed up the process but also help to reduce or eradicate weed seeds and other pests, as well as potential pathogens. In a domestic compost heap, the process is usually less effective, but still adequate to convert garden waste and potentially smelly kitchen waste into a rich, dark material that plants love. In my compost heap, the only two common items from the kitchen that do not get degraded are sweet corn cobs and teaspoons!

In the UK, local authorities are increasingly collecting domestic food waste separately from other rubbish. Instead of being composted, this material can be treated in a more efficient way using anaerobic digestion. The food waste, together with similar material from abattoirs, factories, shops and restaurants, is ground to a slurry, mixed with the appropriate bacteria and put into large tanks. The bacteria, working anaerobically (without air), produce methane as a digestion product. This can be collected and used to produce the power needed to run the plant, with excess electricity being fed into the National Grid. After a few days, the process is complete. The liquid residue can be used as a fertilizer for farms, and the sediment sold as soil conditioner. Anaerobic digesters are used in a similar way in the treatment of sewage (Chapter 9).

Among the negative aspects of microbial biodegradation, the rotting of wood is most important. A fence post is subject to fungal attack in much the same way as any fallen timber, except that it is upright and embedded in the soil. The portion around the soil surface is therefore wet more or less continuously and is also in contact with the air, which is perfect for a variety of fungi to colonize and degrade it. Further up, it is dry most of the time, and below the soil it is anaerobic, so these parts will rot much more slowly.

Inside the house, things are different, and specific forms of rot can be recognized due to different fungi – commonly characterized as either wet rot or dry rot (although both terms are imprecise and rather misleading). The most common form of wet rot is due to the cellar fungus, which is found in timbers in contact with damp masonry, such as that caused by rising damp. A less common form of wet rot is the mine fungus, which causes the wood to shrink and crack into squares – cuboidal cracking – which can be mistaken for dry rot (see below). This is also classified as a brown rot because (like dry rot) it causes darkening of the wood. In contrast, *Phellinus contiguus*, which is a common cause of wet rot of windows and door frames, is a white rot. The many different fungi causing wet rots differ in the extent to which the wood has to be wet, and some will only cause problems with really wet wood, such as that found underneath a leaking kitchen sink.

Dry rot, due to the fungus *Serpula lacrymans*, is much more insidious. It gets its name from the dry, crumbling appearance of the affected wood, not from any ability to attack dry wood. It actually needs a high moisture content to get started – about 30–40 per cent (timber, in a well ventilated area, should have a moisture content of 8–16 per cent). However, the fungus will remain viable with a lower moisture content (down to perhaps 20 per cent). The fungus digests the cellulose within the wood, causing it to darken and to shrink, resulting in cuboidal cracking of the wood. The spores are everywhere, especially if the fungus has been allowed to form mature fruiting bodies.

The best way of preventing dry rot is by maintaining good ventilation, which keeps down the moisture content of the wood. This fungus is notorious for its ability to spread to adjacent areas, even through brick walls – although it needs both the brickwork and the timber on the other side to be damp. The ability to spread in this way makes it extremely difficult and costly to eliminate once it has obtained a foothold.

Timber inside buildings, including furniture, is also subject to non-microbial problems, such as woodworm and death-watch beetle, which are not within the scope of this book.

We can now turn to rock surfaces, including gravestones and stone walls. This might seem to be an unlikely place to find microbes. There are few nutrients available, and the temperature may fluctuate between baking hot and freezing cold, as well as alternating between very wet and very dry. Yet, this is the place where we commonly find one of the most unusual forms of life – the lichens.

Lichens are not actually single organisms, but a symbiotic relationship between a fungus and a photosynthetic microbe (cyanobacterium or alga). The photosynthetic partner contributes to the arrangement by fixing carbon dioxide from the air. The fungus uses the organic nutrients made in this way and, in turn, contributes by using its network of filaments to gather in what water and inorganic nutrients it can find. It should be noted that this is not just a random association. Each lichen consists of an association between a particular fungus and a specific cyanobacterium or alga, to the extent that different lichens are given taxonomic names as though they were single organisms. Nor is the structure of the

Figure 6.6 A selection of lichens on stone and wood.

association random, but rather the photosynthetic partner is located at specific positions within the network of fungal filaments.

Lichens of one sort or another are found all over the world, often in the most unlikely habitats – ranging from polar regions and high mountains to the hottest deserts. Although, in the UK, we are mostly familiar with lichens on the surface of rocks and gravestones – typically as flat, tightly-adhering structures that look merely like discoloured patches of stone – they also occur in a range of other sites, such as on the bark of trees. In tropical rainforests, they can form massive structures hanging from the trees.

Lichens are highly susceptible to atmospheric pollution, especially sulphur dioxide, and during the 19th and 20th centuries they disappeared from many parts of industrial urban Britain. In the UK, the Clean Air Acts (1956, 1968) and subsequent additional controls on emissions have led to the gradual re-emergence of lichens in many places. Their presence provides a valuable monitor of air quality.

Lichens are not the only microbes that will colonize rocks. They tend to start the process, but can then be followed by a variety of other microbes. Bacteria, fungi and algae can all be found on rock surfaces, and even within the structure of the rocks, penetrating several centimetres deep within pores in the structure. They play an important role in the weathering of rocks and the gradual transformation of stone into soil. As the rock surface deteriorates, mosses, ferns and plants start to take hold, accelerating the process. This applies not only to natural features but also to buildings and other structures, where microbial colonization contributes to the discolouration and dissolution of the material. Even concrete is not immune, but can be colonized by fungi and bacteria, which weaken the structure by removing calcium. Although this is a slow process, it is an important consideration if we think about nuclear waste repositories that have to remain intact over many thousands of years. Concrete radioactive barriers in the Chernobyl reactor have been found to be colonized by a variety of fungi.

However, this is a two-way process. Microbes are also responsible for the formation of some types of rock, especially carbonate deposits such as chalk. Various microbes can deposit calcium carbonate externally, while others, such as algae and coccolithophores, use it for cell surface structures. The White Cliffs of Dover are a monument to such microbial activity. It is difficult to imagine microscopic organisms producing such large structures, but we have to remember both the enormous numbers of microbes involved and the extremely long timescale (in human terms) over which they were laid down.

Metal surfaces are also subject to microbial attack (bio-corrosion). We are familiar with the rusting of iron, which is due to chemical oxidation, but corrosion of metals is not always just a chemical process. Microbes are also quite capable of causing corrosion of metals, ranging from ships to water pipes. In the absence of air, bacteria known as sulphate-reducing bacteria (SRBs) produce hydrogen sulphide, which will eat away at the metal. The bacteria are attached to the surface of the metal, forming a biofilm (see Chapter 8), and the reaction occurs locally, just where the bacteria are, so pits form in the metal surface. In the presence of air, there can also be an electrolytic process. Underneath the bacterial colony, electrons are transferred from the bacteria to the metal, which therefore acts locally as an anode. Away from the bacterial colony, the metal acts as a cathode and releases electrons into the water. The (tiny) current that is set up tends to dissolve the metal, causing a pit underneath the colony.

6.5   Extreme environments

Many microbiologists, especially those who concentrate on medically important bacteria, tend to regard the optimum conditions for microbial growth as a temperature of 37 °C, pH about neutral or slightly alkaline (say, 7.5), aerobic, and a medium with plenty of nutrients. For many, or even most, microbes, this would actually be a highly damaging environment. For example, oxygen is a reactive gas and is toxic to cells that have not evolved ways of dealing with it. Also, many microbes will only grow at temperatures that are much lower, or in some cases much higher, than 37 °C. Our human-centred attitude is reflected in the designation of environments that are widely different from the above as 'extreme environments'. To a microbe that lives and thrives in such conditions, other environments might be considered as extreme. However, the term is so deep-rooted that I will carry on using it.

We have already encountered one example – the microbial communities that live in the vicinity of thermal vents ('black smokers') in the deep oceans. These microbes, predominantly Archaea, often have optimum growth temperatures above 100 °C. Other thermophilic organisms are found in natural hot water springs. Although these geysers are hot, the temperature is considerably lower than that in deep sea vents. In hot springs that are not acidic, photosynthetic bacteria (including cyanobacteria) are common and may form green microbial mats. In acidic springs, with a pH that may go as low as pH 1, other bacteria predominate, some of them deriving their energy from oxidizing sulphur or sulphides to sulphuric acid.

Another hot environment is found in oil wells, which may be as deep as 4 km, and where geothermal heating can produce temperatures in the 60–130 °C range. In this anaerobic environment, microbial reduction of sulphate to hydrogen sulphide can cause corrosion problems, as well as reducing the quality of the oil. At the other extreme, microbes are found in deep ice cores from the Antarctic. Although metabolism is extremely slow at such temperatures, these organisms remain viable.

Another extreme environment is exemplified by salt lakes such as the Great Salt Lake (Utah, USA). These may have salt (sodium chloride) concentrations approaching saturation, which would be extremely damaging to most forms of life. However, specially adapted microbes, both bacteria and Archaea (the latter predominate at the very highest salt levels) can thrive under these conditions. These organisms, such as *Halobacterium*, tend to be quite unusual – large, irregular in shape and often with gas-filled vacuoles which may help to keep them near the surface and thus gain access to oxygen, which is poorly soluble under these conditions.

Salt lakes typically form by evaporation of water from lakes with no natural outlet, and therefore the composition of different salts can vary considerably, depending on the make-up of the water that feeds them. The Great Salt Lake has a salt composition similar to that of seawater (although, of course, much more concentrated). Another well-known salt lake, the Dead Sea, is quite different in its composition. It has a high level of magnesium ions, which are even more toxic than the sodium ions in seawater or in the Great Salt Lake. However, it is not 'dead' at all, but contains a variety of microbes, especially Archaea.

The ecology of the Dead Sea is quite complex. It is much deeper (about 340 m) than the Great Salt Lake (which is about 10 m deep), and it contains layers with different salt concentrations, especially after heavy rain, which can produce a more dilute surface layer in which a different spectrum of organisms will proliferate temporarily until the water levels get mixed again. In the Antarctic, lakes with high salt concentration (such as Deep Lake) do not freeze, despite surface temperatures down to $-20\,^\circ$C, because the high salt concentration acts as a natural antifreeze. Halophilic, psychrophilic (cold-loving) microbes have been found in such lakes.

Soda lakes, such as those occurring in the East African Rift Valley, are similar to salt lakes in having a high salt level (ranging from five per cent to saturation) but, in this case, the high salt concentration is combined with extreme alkalinity (pH up to 11). These conditions provide a markedly productive environment, primarily due to the high availability of bicarbonate for photosynthetic production. In the more dilute lakes, with salt (NaCl) concentrations up to 15 per cent, cyanobacteria produce frequent or permanent blooms of surface growth, which can support enormous bird populations, especially flamingos. Lake Natura has an estimated population of two million birds, and they consume some 200 tonnes of cyanobacteria per day.

In contrast to these extremely alkaline environments, highly acidic environments (pH < 3) are not common in nature, apart from acidic peat and the hot springs I referred to earlier. The most important examples of highly acidic environments are those that happen as a consequence of human activity – in particular, drainage water from ore mines and spoil heaps. These typically contain sulphur compounds, which become oxidized to sulphuric acid. Among the bacteria that can produce and tolerate these conditions, *Thiobacillus,* with an optimum pH of 2, is especially common. Microbes growing under these conditions not only have to tolerate low pH, but also high levels of toxic metal ions such as copper and lead, as well as incidental toxic material such as arsenic. I will come back to mine wastes and related topics in Chapter 9.

In this chapter, we have seen something of the amazing diversity of microbes that surround us, as well as being given a taste of their significance (for good or bad) in the environment. In the next chapter, I want to look at how our knowledge of the genetics and molecular biology of microbes (especially bacteria) helps us to understand how bacteria respond to their environment, as well as looking further at some of the exciting developments that have led to an enhanced appreciation of microbial diversity and evolution.

7
Microbial Evolution – Genes and Genomes

Up to this point, I have been largely describing the role of microbes in various contexts – health and disease, food, the environment. This chapter represents something of a change of mood, in looking at how microbes work. The study of bacterial genetics, and the development of the methods known as gene cloning or genetic manipulation, have had repercussions throughout biology. These issues are now so central to the study of the subject that they cannot be avoided (and, indeed, several aspects of this chapter have been referred to already in previous chapters), so the time has come to look at them more systematically. I am afraid this does mean becoming rather more technical than previous chapters so far, but I will do my best to keep it accessible.

7.1 Evolution and inheritance

The concept of species developing through evolution, as commonly ascribed to Charles Darwin (although much modified since), is now completely accepted in scientific circles. Despite that, it still remains a contentious issue, and I will return to the controversy in Chapter 10. For the moment, we can summarize the position as follows.

Variation between individuals exists. In a specific environment, some individuals will be 'fitter' than others – that is, they will grow better, or have more offspring, and so will tend to outperform the others. Consequently, the population of that species, in that specific niche, will become better adapted to that environment. However, in a different niche, the environment will be different, so the variations that result in increased fitness will be different. If these variants can interbreed, their different characteristics will get mixed together – a process known as recombination – so they

---

Understanding Microbes: An Introduction to a Small World, First Edition. Jeremy W. Dale.
© 2013 John Wiley & Sons, Ltd. Published 2013 by John Wiley & Sons, Ltd.

are not likely to develop into different species, although the overall nature of the population might gradually change.

This recombination occurs in two ways. Firstly, in a sexually-reproducing organism, the process of sexual fusion produces a cell with twice the normal number of chromosomes. A process called meiosis then separates these chromosomes, resulting once more in cells with the normal number. During meiosis, the chromosomes are mixed, so the resultant cells will get some chromosomes from one parent and some from the other.

Secondly, at a much lower frequency, there is recombination at the molecular level. This depends on the tendency of similar bits of DNA to pair together and get 'swapped' by enzymes that break and re-join the DNA. The combined effect is that the offspring have a mixture of characteristics of the two parents, and all the variations that have developed become mixed, so there will still be a single 'species' showing a range of variation.

On the other hand, if there are two populations that cannot interbreed, they will gradually become so different that we call them different 'species'. Being unable to interbreed could be because they are separated geographically (e.g. on different islands), but it can also happen for other reasons. For example, two varieties of a plant species which flower at different times will not pollinate one another.

Our definition of 'species' is, to some extent, arbitrary. At one level, we call animals or plants different species because we can recognize them as different. The most inexpert birdwatcher can distinguish a blue tit from a great tit, but even an expert would usually have to rely on song to tell a willow warbler from a chiffchaff. On the other hand, a dachshund and a great Dane are very different, but we would call them both dogs (although, if we had never seen a dog before, we might think they were different species).

Another common way of looking at this is to say that plants or animals are different species if they do not interbreed, or at least cannot produce fertile offspring. Of course, there is some circularity to the argument here, as we have just seen that the lack of interbreeding between variants is one of the factors that leads to them becoming identifiably different species. Another problem is that it is not watertight – there *is* some interbreeding between species in the wild. If we look at a flock of mallards (one of the most familiar ducks), it is not uncommon to see one that has markedly different colouring, due to a different species of duck having mated with a female mallard.

The definition of a species becomes even more problematical when we turn to organisms that do not have true sexual reproduction. This includes bacteria and a range of other microbes. Historically, bacteria have been classified into species, and into higher taxonomic units such as genera and families, on the basis of their morphology (shape) and biochemical characteristics; however, it is only with the coming of modern molecular methods that their classification has been put on a truly scientific basis. Nowadays, we would compare the complete genome sequence of individual organisms and call them different species, genera, families and so on, on the basis of the degree of similarity of their genetic make-up. Such methods, now universal for bacteria, are also becoming widely used for more complex organisms, and we will look at them further later on. Even though the molecular characterization is an objective one, it is worth remembering that it is still to an extent arbitrary – the definition of a species will depend on the level of sequence similarity that we specify.

There were, of course, gaps in Darwin's work. He did not know how the variations occurred in the first place. His ideas, in this respect, tended towards what we now refer to as the inheritance of acquired characteristics – in other words, that the variation occurred as a *response* to the environment. This was commonly exemplified by the somewhat facetious argument that giraffes developed long necks through stretching to reach leaves high up in the trees (an idea often referred to as Lamarckism, but which is rather a parody of the views of Jean-Baptiste Lamarck).

We now know that variations occur spontaneously, all the time, largely as a result of 'errors' in the copying of the DNA of an organism. Many of these changes will be damaging – even lethal – so the organism will not survive. Many other changes will have little or no effect, so they may persist and we only find out about them if we sequence the DNA. Very rarely, however, a change may occur that will be potentially useful and, if the environment provides the right selective pressure, those variants will multiply and eventually become predominant.

It is ironic, given the extent to which the study of bacteria has provided the groundwork for our knowledge of the molecular basis of genetics, that bacteria provided the last refuge for the idea of the inheritance of acquired characteristics. If a patient with an infectious disease is treated with antibiotics, it is not uncommon to find that the bacterium becomes resistant to that drug. The same thing can be done in the laboratory by taking a bacterial culture and adding an antibiotic to it. This may kill virtually all the bacteria, but a few may survive and can be shown to have developed resistance. It appears that they have become resistant *because* of exposure to the antibiotic and, indeed, we often refer to the use of antibiotics as having *caused* drug resistance. This is not the case, though. These resistant mutations happen all the time, spontaneously, but it is only if we add an antibiotic that they have any advantage. In fact, in the absence of the antibiotic, they usually grow less well than the sensitive bacteria without the mutation. Thus, the use of the drug does not cause the mutation – it provides the selective pressure that enables it to dominate.

This can be shown easily in the lab using a technique known as replica plating (Figure 7.1). An agar plate (without antibiotics) is inoculated with the culture so that, after incubation, there is a pattern of colonies on the plate, each colony being descended from a single cell in the original inoculum. A sterile velvet pad is then pressed onto the surface so that a small part of each colony sticks to the pad. This is used to inoculate a fresh plate, with the antibiotic in it, to see whether any colonies grow. This plate now has a replica of the pattern on the original plate so that, if a colony does grow, it is easy to tell which of the original colonies it came from. That

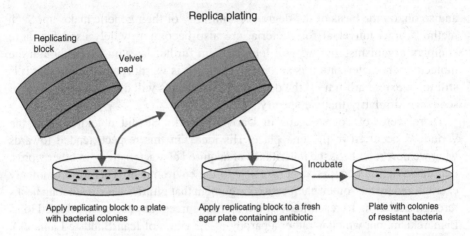

Figure 7.1   Replica plating. The diagram shows how to transfer a small portion of all the colonies on a plate to a fresh plate – in this case, one containing an antibiotic. Only those colonies that are resistant to the drug will be able to grow. By matching the new plate to the original, it is possible to pick off the corresponding colonies and show that they were already resistant even before exposure to the antibiotic. Note that the proportion of resistant colonies would be much less than that shown, so it would be necessary to start with a plate containing many more colonies.

means it is possible to go back to the original plate, pick that colony and test whether it was resistant or not. This will show that the original colony was in fact resistant, even though it had never been exposed to the drug in question.

The reason that this new feature appears to have developed because of exposure to the selective conditions simply reflects the very large number of bacteria involved, and the rapid growth of the organism. Remember that a bacterium like *E. coli* can divide (in the laboratory) every 20 minutes or so; thus, after ten hours, it will have undergone 30 cycles of replication, doubling each time, producing 2^{30} cells – in other words, about 10^9 bacteria (one thousand million bacteria). The mutation may happen very rarely but, even if it occurs in only one in a hundred million cells, these will be ten cells with that mutation by the next day. Adding the antibiotic will kill almost all the cells, but the few that survive will grow rapidly and, after another ten hours, the culture will consist entirely of antibiotic-resistant bacteria.

7.2   Horizontal gene transfer

In general, when we think about genetic inheritance, we mean genes being passed from parents to offspring. This is known as *vertical* transmission of genes. For bacteria, and any other organisms that reproduce asexually, there is only one 'parent', so reproduction leads to offspring that are all are genetically identical (apart from occasional changes that occur spontaneously). However, bacteria are much more versatile than that, in that they can also acquire genes in a way that is not linked to reproduction. They are sometimes able to transfer genes from one cell to another. This happens most often within a single, or closely related, species, but sometimes it occurs across species, even quite different ones. We refer to this as *horizontal* gene transfer, and it plays a major role in the evolution of bacteria, most notably in the spread of drug resistance. It is therefore useful to understand how this works.

Bacteria, in general, have a single piece of DNA, which we refer to as the chromosome (although it is actually quite different in structure from the chromosomes that are found in higher organisms). This carries all the basic information needed for the normal growth and reproduction of that cell. As mentioned in Chapter 4, however, they may also have one or more additional DNA molecules which carry supplementary genetic information; these are the plasmids, which play a key role in horizontal gene transfer – especially in the spread of drug resistance. Plasmids are usually quite small, carrying maybe only a few genes, and they are not essential for normal growth, but the extra information that they carry (including, but not confined to, drug resistance) can greatly expand the capability of the species as a whole.

Plasmids were first recognized in *Staph. aureus* strains that were resistant to penicillin. These plasmids carry a gene that codes for an enzyme (β-lactamase, or penicillinase) that destroys the antibiotic (see Chapter 4). Sensitive strains did not have this plasmid, and resistant ones that lost the plasmid became sensitive.

Around the same time (in the 1960s), an interesting phenomenon was observed in Japan, in the bacteria that cause dysentery (*Shigella*). From some patients with dysentery who were treated with an antibiotic, resistant bacteria were recovered. These bacteria had developed resistance not only to the antibiotic that the patient had been treated with, but to several other unrelated drugs as well. This was entirely unexpected, as there was no selective pressure for these extra resistances. Furthermore, it was found that if one of the multi-resistant strains was mixed with a sensitive one, the latter strain could become resistant to all of the drugs that the original resistant isolate could tolerate.

It was eventually realized that this occurred because all of the genes involved were carried by a single plasmid, so selecting for one resistance selected for all of them. Furthermore, this plasmid had the ability to transfer from one bacterium to another, and not just to other *Shigella* strains but also to related bacteria, such as *Salmonella* and *E. coli*. The reason that the initial *Shigella* strain became resistant was that it had acquired a plasmid from other bacteria that were already present in the gut of the patient.

Transfer of a plasmid from one cell to another occurs by a process known as *conjugation*. Some plasmid genes code for long, thin structures (pili) that stick out from the surface of the cell (these are often referred to as 'hair-like' but they are not really like hairs, apart from being long and thin). In a mixture of cells, these pili latch onto the surface of other cells and then contract to bring the two cells together. The plasmid is then transferred from one cell to the other, being copied in the process – so we should really refer to transfer of *a copy* of the plasmid.

Antibiotic resistance plasmids, which can carry several different resistance genes and can transfer themselves from one bacterial cell to another, have played a major role in the spread of resistance, not just within one species but across wide taxonomic boundaries. There are endless varieties of different plasmids, carrying various combinations of resistance genes (and other characteristics), and they can change by acquiring new genes or losing them. How do these plasmids acquire this repertoire of genes, and what is the basis of this fluidity in their makeup? Our understanding of this developed from the identification of special mobile genetic elements called *transposons*.

This discovery derived from laboratory studies with a strain of *E. coli* that carried two plasmids – one that had a gene conferring resistance to ampicillin (an antibiotic of the penicillin group) and a second that had a streptomycin resistance gene. When this strain (the *donor*) was mixed with another *E. coli* strain (sensitive to both drugs), it was found that the second strain (the *recipient*) became resistant to both antibiotics at a much higher frequency than would be expected for independent transfer of both plasmids.

Plasmid transfer in such a system is often quite a rare event. If one in a million cells acquire a resistance plasmid, this is quite easily detected by growing the cells in the presence of that drug. However, if the two plasmids are transferred independently, that would mean that only one in a million million cells – 1 in $(10^6 \times 10^6)$ cells or 1 in 10^{12} cells – would have acquired both plasmids, which would be practically impossible to detect. It was therefore surprising to find detectable levels of transfer of both resistance genes. Even more surprising was to find that the newly resistant recipient cells had only a single plasmid. The explanation was that the ampicillin-resistance gene had 'jumped' from one plasmid to the other – it had *transposed*. From that, the DNA element that was capable of this behaviour was termed a *transposon*.

There is a wide variety of transposons, and they vary considerably in their structure, the genes that they carry and the way in which they transpose. The simplest consist of just two genes – a transposase, which is responsible for its mobility, and an antibiotic resistance gene. The transposase breaks the DNA at each side of the transposon and joins it to a different bit of DNA, either on another plasmid or on the chromosome, which moves the transposon (or a copy of it) from one site to another.

Transposons can also acquire additional genes from the bacterium they happen to be in. This includes antibiotic resistance genes, which are quite common in environmental bacteria. Remember that antibiotics are originally natural products, produced by microbes. They perhaps have a survival value for those organisms in that they are able to kill other bacteria that might otherwise use scarce resources, e.g. in the soil. Why are the producing organisms not killed by their own antibiotics? Because they carry genes that make them resistant.

There are also similar mobile pieces of DNA that do not carry any genes at all, other than the transposase needed for their transposition. These are known as *insertion sequences*. As they do not carry any genes, other than those needed for their own mobility, they do not add any new features to the host bacteria, but they were recognized a long time ago because of their ability to inactivate genes. When an insertion sequence jumps into a gene, it will inactivate that gene, thus causing a mutation. Such mutations are quite common, as most bacteria carry many copies of a number of different insertion sequences, as is now recognized as a result of genome sequencing. These do not contribute to the fitness of the cell, but they survive in the way any other well-adapted parasite survives – by causing as little damage as possible and by possessing mechanisms that ensure their survival, in this case by transposition. If a copy of an insertion sequence is lost, then one of the remaining copies will transpose and replace it.

So we now have plasmids that can move from one bacterium to another, transposons that can move between two plasmids (or between plasmids and the chromosome) in any cell, and insertion sequences hopping around inactivating genes – a picture of the genetic make-up of a bacterium that is much more fluid than just a single chromosome inherited from the parent cell. But that is still not the end of the story.

Apart from antibiotic resistance, plasmids can carry many other sorts of genes, notably those responsible for the virulence of a pathogen. For example, the toxin that is responsible for the symptoms of tetanus, caused by *Clostridium tetani*, is due to a plasmid-borne gene. Bacteriophages (viruses that infect bacteria) can also carry virulence genes. For example, only those strains of *Corynebacterium diphtheriae* that carry a specific phage are able to cause diphtheria, and similarly scarlet fever is caused by strains of *Streptococcus pyogenes* that carry a specific phage. In both cases, the toxin that is responsible for the symptoms of disease is coded for by a gene carried by the phage. The ability of phages to transfer genes from one bacterium to another is referred to as *transduction*.

We cannot be sure how these phages acquired the toxin genes, nor from what source. But we do know that some phages can acquire genes from the chromosome when they are able to establish a permanent presence in their host rather than killing it. The phage remains within the bacterial cell, integrated into the chromosome, as a stable parasite. This state is known as *lysogeny*. It can remain in this state indefinitely, but sometimes things go wrong and the phage DNA is cut out of the chromosome and starts replicating. If an error is made in cutting the phage DNA out of the chromosome, some adjacent genes are incorporated into the phage DNA, thus leading to those genes being transferred to a new bacterial host when the phage infects it.

Genome sequencing (see later in this chapter) has shown that most bacteria carry a substantial number of integrated phages, many of which show no signs of being active viruses. We can recognize their presence because their genes are related to those of other known viruses. We can also recognize these regions of the bacterial chromosomes as being in a sense 'foreign', because they differ from the rest of the chromosome, especially in the distribution of the bases (A, G, C, and T). Any species has a characteristic ratio of G + C to A + T, known as the GC content, which is (within limits) constant throughout the chromosome. However, where regions of DNA have originated from a different species, there will be a sudden change in the GC content as we read along the genome sequence. Since the most interesting of these different regions are associated with genes that are involved in pathogenicity, they are often known as pathogenicity islands – but other properties are also involved, so I will just refer to them as *islands*.

In many cases, these islands accompany identifiable phage genes, indicating the presence of viruses in the bacterial chromosome; in some, these are responsible for the ability of the bacterium to cause disease. One of the most important of these is in the bacterium that causes cholera (*Vibrio cholerae*), where the toxin gene that is responsible for the symptoms has been inserted into the chromosome by a bacteriophage.

So, bacterial cells can acquire DNA from other sources by cell to cell contact (conjugation), or by phage-mediated processes (transduction). There is one further avenue open to them, which is to take up DNA directly from its environment; this is known as *transformation*.

One example of the consequence of this process comes from the bacterium *Streptococcus pneumoniae*, familiarly known as the pneumococcus. This organism is not only one of the main causes of bacterial pneumonia but also can cause meningitis (see Chapter 3), so is a serious pathogen. The pneumococcus can develop penicillin resistance by producing a changed version of a protein called penicillin-binding protein, which is the target for the action of penicillin. The sequence of the gene tells us that much of the protein is the same as in the sensitive strains – apart from one region, which is the key part for the action of penicillin. This region resembles part of a gene from streptococci that are normal inhabitants of the mouth, which are intrinsically more resistant to penicillin than the pneumococcus is. Thus, it seems that the pneumococcus has developed resistance by transformation with DNA from an oral streptococcus – but only a part of the gene has been affected.

Something similar was responsible for the development of methicillin-resistant *Staph. aureus* (MRSA), which owe its resistance to methicillin to the presence of a gene coding for a new penicillin-binding protein. It is not just a mutation of the existing gene; sequence data tells us that it is substantially different and must have originated elsewhere. In this case, we do not know the origin of the gene, but it is likely to have come by transformation with DNA from another species.

Transformation was first discovered in 1928 by Fred Griffith, working with the pneumococcus. This bacterium is only able to cause disease if it has a polysaccharide capsule which protects it against the body's defences (see Chapter 3). When it is grown in the laboratory, it loses this ability and this can be easily recognized by a change in the appearance of the colonies on an agar plate. Capsulated (virulent) bacteria produce neat smooth colonies but, when they lose the ability to make the capsule, the colonies are rough and untidy looking.

Fred Griffith found that if live rough, non-virulent, cells were mixed with a heat-killed extract of a virulent (smooth) strain, they became virulent again – they produced smooth colonies and were able to kill mice. Something in the heat-killed extracts, which he called the *transforming principle*, had transformed the non-virulent cells into virulent ones.

As a digression, it is worth pointing out that this was a landmark observation, because it enabled the identification of DNA as the genetic material of the cell. Extensive purification of the transforming principle enabled Oswald Avery, Colin Macleod and Maclyn McCarty, at the Rockefeller Institute for Medical Research in New York City, in 1944, to show that when all other material, apart from DNA, had been removed, it was still able to transform pneumococcal cells. This was surprising at the time, and not fully accepted immediately.

DNA had been known for the best part of a century and had been thought to be merely some sort of structural material. After all, DNA is superficially a very simple structure, consisting merely of a sugar (deoxyribose), phosphate and four bases (adenine, guanine, thymine and cytosine). How could such a simple chemical carry all the information needed for the synthesis of proteins, which are made up of 20 different amino acids? Eventually it was realized that reading a four-letter code in groups of three (*codons*) produced 64 (2^4) 'words' – more than enough to code for the sequence of amino acids in a protein. Although this is now so firmly established that it barely needs saying, I remember as an undergraduate having a series of lectures on the evidence for the triplet code.

Apart from its significance in the development of molecular biology, transformation is important in bacterial evolution. Although it has only been demonstrated in the laboratory in a limited number of species, this is thought to be because it is too rare an event. Out there in the wild, however, the number of organisms is far greater than we can handle in the lab and the timescale is much larger than we would have patience, or funding, for – so there is plenty of opportunity for extremely rare events to have a significant impact.

We can see, or at least infer, the consequences of this. In addition to the examples above, we can consider *Neisseria gonorrhoeae* (the gonococcus, which causes gonorrhoea), which attaches to human cells by means of surface appendages (pili). Antibodies to those pili prevent attachment, but the pili are highly variable, and thus the bacteria can escape the antibodies. There are two reasons behind this variability. The first is that the gonococcus has many versions of the pilin gene, although only one is expressed at a time. If a copy of one of the non-expressed genes replaces the expressed gene, then a different pilus is made (we will encounter a similar process later on, in the antigenic variability of the trypanosome that causes sleeping sickness). However, they are not satisfied with only that much variation; they can also acquire bits of DNA from other strains they encounter, by transformation, so adding considerably to their repertoire. Furthermore, these DNA movements do not always involve whole genes, but can mix and match parts of genes, thus giving an enormous range of possibilities.

We have now built up a picture of the genetic make-up of a bacterium, not just as a single chromosome subject to minor changes by mutation but as a highly fluid feature, with additional genes contributed by plasmids and bacteriophages, with bits of DNA coming from various sources by different mechanisms and with transposons and insertion sequences moving around the chromosome and between the chromosome and plasmids. The advent of genome sequencing, and the now-routine sequencing of the genomes of many varieties of the same bacterial species, are throwing up more and more examples of this fluidity of their genetic make-up. The extent to which this occurs is remarkable; genome sequencing of *E. coli* strains has shown that only 40 per cent of genes are present in all strains. In addition to this inherited fluidity, there is another level of variation, in that bacteria respond to changes in their environment by switching genes on and off.

7.3   Variation in gene expression

A gene is of no use to a cell unless it is expressed. This means that the protein for which it carries the genetic information has to be produced. There are two stages to this process: first, the DNA has to be transcribed into RNA, using RNA polymerase; second, ribosomes have to translate that RNA into protein. These processes were summarized in Chapter 1.

For RNA polymerase to transcribe a gene, it must recognize a specific site known as a promoter. This is where it binds to the DNA, next to the gene in question, and starts making RNA, using the information in the DNA to determine the sequence of bases to be incorporated. One of the simplest of control mechanisms involves another protein (a repressor) binding to the DNA and preventing the RNA polymerase from attaching to the promoter. For example, *E. coli* will only make the enzyme β-galactosidase, which is needed for breaking down lactose, if lactose is available. In the absence of lactose, a repressor protein binds to the DNA near the promoter site (see Figure 7.2) and prevents the RNA polymerase from transcribing that gene (i.e. copying it into RNA). However, when lactose is added, it *induces* expression of β-galactosidase. Binding of inducer to the repressor changes the shape of the repressor protein, which prevents the repressor binding to the DNA, so that the gene can be expressed.

Another way in which genes can be switched on or off concerns the nature of the promoter. There are several different classes of promoter, and normally RNA polymerase will only recognize one type. Therefore, when the bacterium is growing normally (e.g. *E. coli* growing in the presence of air and plenty of nutrients), some genes are expressed, while others that are not needed under these conditions are turned off. They have different promoters, which are not recognized.

At the end of active growth, when nutrients start to run out, the bacterium enters a different phase of 'growth' known as stationary phase. Although there is no net growth, the cells are not inactive and they need a number of functions in order to be able to survive. Thus, the genes that were needed for active growth are switched off

REGULATION OF β-GALACTOSIDASE IN *E. COLI*

(a) No lactose in growth medium; β-galactosidase not needed

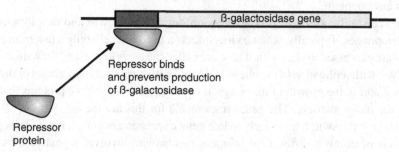

(b) Lactose (inducer) present

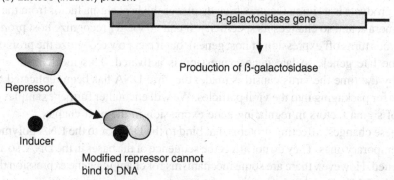

Figure 7.2 Regulation of gene expression. (a) If there is no lactose available, the bacterium does not need to make β-galactosidase. A repressor protein binds to the DNA and prevents expression of the gene. (b) When lactose is added, an inducer binds to the repressor and changes its shape so that it can no longer bind to the DNA. As a result, the gene is expressed and β-galactosidase is made.

and a different set of genes is activated. This is due to a change in the RNA polymerase that enables it to recognize a different class of promoter and, hence, express a different set of genes. Technically, this change occurs through the action of another protein known as a sigma factor, which binds temporarily to the RNA polymerase. These sigma factors determine the specificity of the RNA polymerase, so, during active growth, one sigma factor binds to the enzyme and enables it to recognize those promoters. When the bacterium goes into stationary phase, that sigma factor disappears and is replaced by a different one. The RNA polymerase now recognizes a different set of promoters and so transcribes a different set of genes.

Similar switches occur under other conditions – for example, if the bacteria run out of air (they start growing anaerobically) or if they are exposed to certain hostile conditions, such as high temperatures. Different sigma factors are activated, the specificity of the RNA polymerase changes and there is a major shift in the set of genes that are expressed. The bacterium thus has a number of means available by which it can switch genes on or off, individually or in blocks, in response to changes in its environment.

We see similar shifts in gene expression during the growth and development of bacteriophages. Typically, when a virus infects a bacterial cell, only a few of the viral genes are expressed to start with. These are called the *early* genes, and they are usually involved with replication of the phage DNA and with turning off expression of the host genes. Later in the growth of the phage, it needs to start making the proteins that will form the phage particle. The genes responsible for this are the *late* genes.

Although the switch from early to late gene expression occurs in most viruses, the methods of doing it differ. One frequent mechanism involves sigma factors. The early genes can be expressed using the host RNA polymerase, unmodified, as they have promoters that are recognized by the host enzyme. However, one of the early gene products is a sigma factor, which displaces the host sigma factor from the RNA polymerase and so changes its specificity. It can no longer recognize host promoters (and thus turns off expression of host genes), but it can now recognize the promoter(s) for the late genes, so late gene expression is activated. This mechanism ensures that by the time the virus capsid is made, the viral DNA has been replicated and is ready for packaging into the viral particles. We will encounter further examples of the use of sigma factors in regulating gene expression in the next chapter.

These changes, affecting proteins that bind to the DNA or to the RNA polymerase, are temporary ones. They do not affect the sequence of the bases in the DNA, so are not inherited. However, there are some mechanisms for changing gene expression that do alter the DNA. For one example we can turn to a protozoan parasite, *Trypanosoma brucei*, the cause of sleeping sickness, which was introduced in Chapter 4.

This organism is transmitted by the tsetse fly. When someone is bitten by an infected tsetse fly, the protozoa enter the bloodstream and multiply there. As the numbers build up, the first symptoms develop, mainly an acute fever. This lasts for a day or two before subsiding, as the body starts to produce antibodies which virtually eliminate the parasite. However, it has not disappeared completely, and soon there is a second wave of fever as the numbers build up again. Once more it subsides as the antibodies respond to it. This cycle of periodic bouts of fever continues, over and over again. This is in itself quite debilitating, but the most serious consequences arise when the parasite starts to affect the brain. At this stage, the patient lapses into a coma (hence the name sleeping sickness) and will eventually die. The relevant factor here is, why do these periodic bouts of fever happen? Why do the antibodies not eliminate the parasite completely?

The antibody response is directed at a molecule on the surface of the parasite, known as the Variant Specific Glycoprotein, or VSG (a glycoprotein is a protein with sugar molecules attached to it). After the antibody response has eliminated most of the parasites, the VSG changes and these antibodies are no longer effective, so there is a second wave of fever. Antibodies to the new VSG are then made, the parasite declines and the VSG changes again. And so on.

You might think that this is due to a mutation in the gene responsible, but that is not the case. At the start, the trypanosomes are capable of producing a wide range of VSGs, but only one form is actually made. All the genes are there, but only one version is expressed. The rest are maintained in 'silent' sites on the DNA. For a new VSG to be expressed, that gene moves from a silent site to an active one, replacing the copy that is already there. The organism thus moves systematically through a repertoire of over 1,000 different VSG genes, driven by the selective pressure of the antibody response eliminating all the parasites that express the previous copy. Compare the description of antigenic variation in the gonococcus, described earlier.

A superficially similar example comes from a bacterium. Species of *Salmonella* that cause food poisoning produce a surface appendage known as a flagellum (which is responsible for movement of the bacteria). The flagellum is also known as the H antigen and, in the diagnostic laboratory, strains are distinguished by the reaction of the H antigen with specific antibodies. However, many strains can produce two different types of flagella, which I will call H1 and H2 – and the bacteria can switch reversibly between them. In this case, the switch does not happen by movement of the genes, but by inversion of a region carrying a promoter. For a promoter to work, it not only needs to be adjacent to the gene that it is controlling, but it also has to be facing the right way. In this case, the promoter is next to a pair of genes that code for the H2 antigen and a repressor of the H1 gene (this repressor binds to the DNA next to the H1 gene and inactivates the promoter of that gene). If the promoter is facing towards those genes, then the bacterium will make the H2 antigen and the repressor of the H1 gene, so the H1 antigen will be switched off by the repressor. If the bit of DNA carrying the promoter flips round so it is facing the wrong way, then neither of these genes will work. The H2 antigen will not be made, nor will the repressor of the H1 antigen, so the H1 antigen is produced instead.

Although not nearly as versatile as the trypanosome system, this has similar advantages for the *Salmonella* in that, by changing its antigenic structure, it can temporarily evade the immune response.

7.4   Gene cloning and sequencing

The techniques of gene cloning and sequencing have caused a revolution in our understanding of all aspects of biology, as well as providing the means for specific manipulation of microbes (and higher organisms). Some aspects have already been referred to in earlier chapters, and more will come later on. In order to understand these developments, I will describe briefly some of the core techniques, while trying to avoid getting into too much technical detail.

To start with, it is worth clarifying the meaning of the word 'cloning'. Clones are individuals that originate by asexual reproduction from a single individual. They are therefore genetically identical, subject to the normal processes of variation. Since bacteria only reproduce asexually, a bacterial culture can be considered as a collection of clones. Furthermore, if we take a mixed culture and spread it on an agar plate so that each individual cell will give rise to a single colony, and we then pick one colony to produce a culture, we have cloned that cell.

With 'higher' organisms, the distinction is between normal reproduction, by sexual means, and asexual reproduction. Although the cloning of animals is a new procedure, still in its infancy, the application of cloning to plants has been around for thousands of years. Any gardener will be familiar with the technique of taking cuttings from a plant to produce a lot of new plants for the garden. This is cloning, as is digging up a daffodil and splitting the collection of bulbs. Cloning plants is therefore merely an extension of a completely natural process, as very many plants reproduce themselves in this way, either partly or entirely.

With gene cloning, we introduce a gene into a cell (a bacterial cell for current purposes) and then produce a culture of that cell. Since every cell now has a copy of the gene we introduced, we say we have cloned that gene. How is this done?

Let us assume we are trying to clone a gene from a bacterium, say *Staphylococcus*, and we want to put it into *E. coli* (the actual techniques are much the same, whatever the source of the DNA). We would take a staphylococcal culture, extract the DNA and break it up into bits (Figure 7.3). This can be done mechanically, by mixing it vigorously, or by treating it with specific enzymes known as restriction enzymes (which I will come to shortly). It is no use just putting those bits into a bacterial cell – they need to be integrated into a plasmid if they are to be copied by the host cell.

The plasmid DNA is a circular molecule, so we have to open it up to insert our foreign DNA. This is where restriction enzymes are really useful. There are lots of different ones, but each is able to cut DNA at a specific site. They recognize short sequences of DNA (usually 4–6 bases) and break the chain at that point. It is quite easy to arrange things so that a specific restriction enzyme will cut the plasmid at a single point only. We then simply add a small amount of enzyme to the plasmid DNA, incubate it for a while, and the job is done.

Now we have to join our foreign DNA to the plasmid. For this, we use another enzyme known as DNA ligase, which joins DNA fragments, end to end. We mix our cut plasmid with the fragments of DNA to be cloned, add a small amount of ligase and incubate it again.

The next step is to get our *E. coli* to take up the DNA. There are different ways of doing this, but one of the simplest is called electroporation. We add the DNA mixture to the *E. coli* cells and subject them to a brief pulse of high-voltage electricity. The mixture is then spread out on an agar plate and we wait for the colonies to grow. The efficiency of uptake of DNA is low, so only a minority of cells will actually acquire a plasmid. This is not a problem, however, as the plasmid has an antibiotic resistance gene and adding that antibiotic to the plate ensures that only cells that have got the plasmid will be able to grow. The low efficiency of DNA uptake is actually an advantage, as it ensures that no cells will have acquired more than one plasmid.

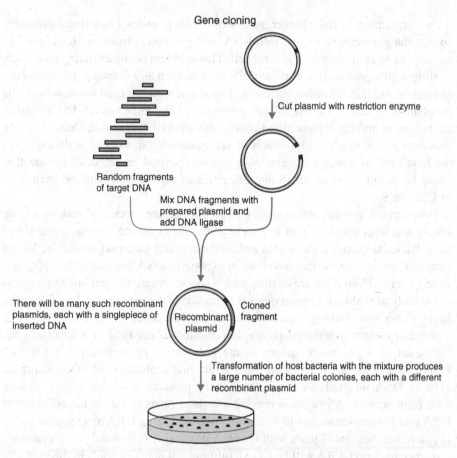

Figure 7.3 Gene cloning. The plasmid is cut with a restriction enzyme, which converts it from a circle to a linear molecule. This is mixed with the DNA fragments to be cloned, and DNA ligase is used to join the molecules together. The resulting mixture of recombinant plasmids is put into a host bacterium by transformation (e.g. by electroporation). Each colony that is formed contains a different recombinant plasmid.

Note that there has been no attempt to separate out the bits of staphylococcal DNA. Each fragment will have been joined to a different plasmid molecule, so each *E. coli* cell will have taken up a different recombinant plasmid – that is a plasmid carrying a bit of foreign DNA. We will thus have a mixture of colonies on the plate, which is known as a gene *library*. The trick is to find which of the cells contains the gene that we are interested in; there are a number of ways of doing this, but at this point things become rather too technical for this book.

Using these techniques, we can (in our example) insert a staphylococcal gene into *E. coli*. However, in order to be able to make full use of this capability, we need to get that gene to work – in other words, it has to be expressed, resulting in the production of the protein that it codes for. This does not necessarily happen by itself.

As we saw earlier in this chapter, gene expression requires a functional promoter so that the gene can be copied into RNA, and promoters from one bacterium do not always work in another bacterial cell. This problem becomes more acute when dealing with a gene from a non-bacterial source, such as a human gene, where the expression signals are rather different. The major strategy used to overcome this problem is to add a known *E. coli* promoter to the gene, which has a further advantage in making it possible to choose the level of expression. Using a strong promoter would result in high levels of expression, but this could be damaging to the host cell, in which case a weaker promoter could be selected, or one that could be switched on or off. Some examples of this at work will be dealt with in Chapter 9.

Nowadays, it is often not necessary to go through the process of making a gene library and screening it to find a specific gene. The complete genome sequence of many bacterial species is known (as well as that of many larger organisms, including humans), and we can use that information to obtain a DNA fragment containing any gene we wish. Even if we are dealing with a species where the genome sequence is not already available, it is often quicker to sequence the genome (as I will describe later), rather than making a gene library to get that gene.

We can use the genome sequence information to get hold of a specific gene by means of a procedure known as the polymerase chain reaction (PCR; see Figure 7.4). This takes advantage of the fact that replication of DNA, using an existing DNA molecule as a template for the production of new copies, cannot start from scratch. A *primer* is needed – a short (usually 20–30 bases) piece of DNA that is complementary to one strand of the existing DNA at a specific point. If you remember that G pairs with C and A pairs with T, then the complementary sequence to (say) GCTA will be CGAT (although it actually reads in the opposite direction – TAGC). The DNA that we start with is double stranded, but the two strands are easily separated by heating them. If the primer is then added and the mixture allowed to cool, then the primer will bind, at a specific site, to the complementary region of the template. The DNA polymerase can then use it as a starting place for the production of a new DNA strand that is complementary to the template.

We can also use a second primer, complementary to a region of the second strand but on the far side of the gene in question, to start the synthesis of another strand, which runs backwards towards, and as far as, the place where the first primer was. The result will be a complete copy of the region between the two primers.

In practice, all these steps can be combined into one process. A mixture is made containing the template DNA, the two primers, a heat-stable DNA polymerase and the substrates needed for DNA synthesis. This mixture is boiled for a minute or two to separate the two strands of the template, then it is allowed to cool so that the primers bind to the template DNA, and the temperature raised to one that allows the DNA polymerase to work. This produces a copy of the region of DNA that needs to be amplified.

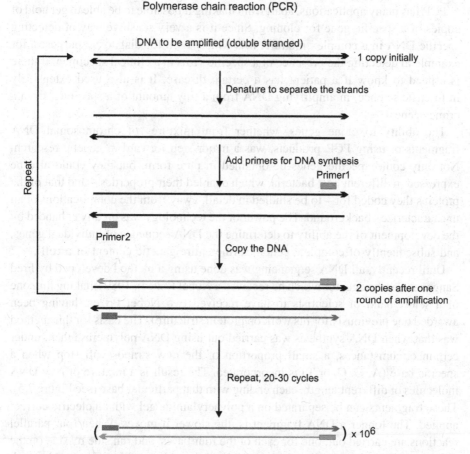

Figure 7.4   Polymerase chain reaction. See the text for explanation. The arrows are added to the DNA strands to show their orientation and the direction in which the new strands are made.

The process is then repeated – boiling, cooling, raising the temperature to that needed for the enzyme reaction – over and over again. Each cycle will double the amount of the DNA region being amplified. After ten cycles, there will be a thousand copies; after 20 cycles, a million copies; after 30 cycles, a thousand million (10^9) copies... Well, at least in theory. After a while, it starts to become less efficient as substrates run out.

However, it means that a lot of specific DNA can be produced, starting with a single molecule. And it is not complicated to carry out – all these cycles of changing the temperature are done by a machine that can be programmed to suit whatever specifications are called for. Of course, the sequence of the DNA will need to be known (or at least an educated guess made) so that suitable primers can be designed, but these are easily synthesized and there are many companies that will tailor them to specific requirements, quite cheaply.

PCR has many applications, apart from making it possible to be able to get hold of copies of a specific gene for cloning. Since it is a very sensitive way of detecting specific DNA in a complex mixture, it is widely used for diagnostic purposes, for example in detecting the presence of a specific virus in a clinical specimen, if there is a need to know if a patient has a certain disease. It is also used extensively in forensic science, in amplifying DNA from a tiny amount of a specimen from a crime scene.

The ability to clone genes, whether from mixtures of chromosomal DNA fragments or using PCR products, was a major step forward in genetic research. Not only could specific genes be obtained in pure form, but they could also be expressed in different host bacteria, which enabled their properties – and that of the proteins they coded for – to be studied in detail, away from the complications of an uncharacterized background. The power of the technology was greatly enhanced by the development of the ability to determine the DNA sequence of individual genes, and subsequently of complete genomes (the entire genetic content of a cell).

Until recently, all DNA sequencing was done using a method developed by Fred Sanger at Cambridge, for which he received a Nobel Prize in 1980 (making him one of a select band of scientists to have received two Nobel Prizes, having been awarded one previously for his work on protein structure). The basis for this method was that when DNA synthesis was carried out using DNA polymerase then, under certain circumstances, a small proportion of the new strands will stop when a specific base (A, G, C, or T) is incorporated. The result is a mixture of new DNA molecules of different length, each ending with that particular base (see Figure 7.5). These fragments can be separated on a polyacrylamide gel with an electric current applied. The longer a DNA fragment is, the slower it moves. When four parallel reactions are carried out, one for each of the four bases, and run side by side on the gel, the result is a ladder of bands that enables the sequence of those bases to be read.

In the early days, when this was done manually, only a sequence of 50–100 bases could be read from a single gel. A typical gene is much longer than this – some 500–1,000 bases or more. However, by doing a series of such reactions using randomly fragmented, overlapping bits of DNA, the data could be pieced together by a computer to obtain longer sequences. Later technical developments, using robotic methods to set up the reactions and machines to read the results, enabled longer sequences to be read, so that sequencing individual genes became straightforward and sequencing longer DNA regions – including whole genomes – became possible.

The first complete genomes sequenced were those of simple bacterial viruses, starting, in 1975, with one called ØX174, which has only about 5,000 bases in its DNA. The first complete sequence of the genome of a cellular organism, that of the bacterium *Haemophilus influenzae* (1.8 million bases), was achieved in 1995. Many other complete genome sequences followed, including the human genome – mainly at specialist centres where batteries of sequencing machines were employed.

However, there are serious limitations to this approach. Firstly, in order to obtain the libraries of small fragments need for sequencing, those bits of DNA had to be

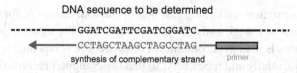

DNA sequence to be determined

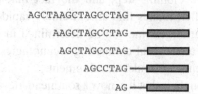

Under specific conditions, a mixture of fragments will be
generated, each ending with an A:

Similar reactions are carried out for G, C and T. The four
sets of fragments are separated on an acrylamide gel

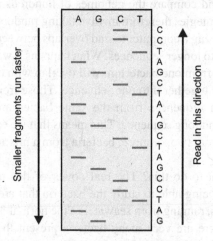

Figure 7.5   Determination of DNA sequence. The diagram shows the basic principle of the Sanger
method. Note that there are several more effective methods available now.

cloned and then each clone had to be grown and its DNA extracted. Even with
automated technology, this took time. Secondly, the piecemeal approach of
sequencing each bit in turn was also time-consuming, even with large numbers
of sequencing machines being used.

In recent years, radically new technologies have been developed to circumvent
these problems. One such method involves attaching the DNA fragments to
microscopic beads and then encapsulating the beads, together with the reagents
needed for a PCR reaction, in oil droplets. The fragments are then amplified by
PCR, so that there are millions of copies of that fragment attached to each bead. The
beads are then loaded into millions of tiny wells (so each well has one bead), each
well being attached to an optic fibre cable. The whole assembly is washed with a

DNA synthesis mixture containing just one of the four bases (A, for example). If a reaction happens in a well, and A is incorporated, a light signal is generated from that well, and this is read by a computer. The procedure is repeated with each of the four bases, sequentially and repeatedly, so that the computer records the sequence in which each base is added in each of the millions of wells.

The absence of a cloning step, and the fact that millions of sequences are determined in parallel, makes this an extremely rapid and efficient system. The sequence of a bacterial genome can be determined in just a few hours. It should be noted that there are several competing technologies, currently, and other even more efficient systems are under development.

The outcome, though, is that it is now a routine matter to sequence the genomes of smaller organisms such as bacteria. Even with larger and more complex genomes, including humans, the timescale, and the cost, have come down to a level where it is possible to sequence and compare the genomes of hundreds of individuals.

These sequencing strategies therefore involve taking random fragments of DNA, sequencing them, and using a computer to find overlaps between the sequences so as to join them together into longer sequences. What happens if we start with a mixture of bacteria? Sequences from one bacterium will overlap with others from the same organism and be joined together into one sequence. Those from another bacterium will only overlap with sequences from the same bacterium, so will be joined together to form a second long sequence. This means that we can get at least part of the genome sequence from two, or more, bacteria from a mixture without having to purify the bacteria.

Why would we want to do this? The real power of this procedure, known as *metagenomics*, lies in being able to study the bacteria that are present in complex environmental mixtures, ranging from seawater to the normal flora of our intestines. In such a situation, there are very many bacteria present that we know little or nothing about, and many of these have not even been grown in the lab. We have already encountered applications of these techniques, in Chapters 2 and 6. If we had to isolate and grow these organisms first, we would only be looking at an unrepresentative sample. However, since we do not need to purify the bacteria, we do not have to be able to grow them in the laboratory. We can rely on the amplifying power of PCR to produce enough DNA for the sequencing reactions, even from the relatively small numbers of some bacteria that may be present in our sample (you may object that I earlier said that we have to know the sequence first in order to be able to amplify a DNA fragment – but that is only true if we need to amplify a *specific* fragment. In this case, we want to amplify all the random fragments, and there are various tricks that can be used to achieve that).

So we can take our environmental sample, isolate all the DNA from all the organisms present, randomly amplify all the fragments and sequence the mixture. This will give a set of partial genome sequences. By comparison of those sequences with the very extensive database of known DNA sequences, we can identify the presence of any known organisms. Experience with these techniques has shown that a high proportion of the information gathered does not correspond to that from any known organism – whether we are considering the human gut or samples of seawater. It is apparent that the diversity of microbial life is far greater than we had previously appreciated.

In summary, the study of the genetics of bacteria (and other microbes) has told us a lot about how they work, as well as providing the ability to manipulate them to do all sorts of useful things. In the next chapter, I will consider some aspects of bacterial communication and development, a field that has benefited greatly from genetic studies, and in Chapter 9 I will look at microbial biotechnology, which includes the exploitation of genetic manipulation.

8

Microbial Development and Communication

Typically, in the laboratory, with a liquid bacterial culture, we could regard each bacterial cell as swimming around freely, independently of any other cells in the culture. And in this situation, the simple picture is that each bacterial cell gradually gets bigger until a critical size is reached, when it divides into two cells. While we can learn a lot about bacterial behaviour from such a simple model, there are many respects in which it is not an adequate reflection of the real situation in nature. The purpose of this chapter is to illustrate some of the ways in which bacteria and other microbes are organized, how complex structures develop, and how cells communicate in multicellular communities.

8.1 Cell division

The simple model, of a bacterial cell increasing in size and then dividing into two cells, poses a number of difficult questions if we look at it carefully. If we consider a basic cell as having a single copy of its DNA, then it is obviously important that, when the cell divides, each of the daughter cells has a copy. Therefore, cell division must be linked to chromosome replication, so that division does not happen until replication is complete. Furthermore, just having two copies of the DNA is not enough. If the two copies were distributed randomly, then, when the cell divided, some of the progeny would have two copies and some would have none (and would die). The answers are elegant, and rather complex.

The obvious model – that there are signals that tie together chromosome replication and cell division – is only partly correct. It is true that cell division will only happen once the DNA has been copied, but the start of DNA replication is not linked directly to cell division. The key fact here is that replication of the chromosomal DNA in *E. coli* takes about 40 minutes, and a further 20 minutes is

Understanding Microbes: An Introduction to a Small World, First Edition. Jeremy W. Dale.
© 2013 John Wiley & Sons, Ltd. Published 2013 by John Wiley & Sons, Ltd.

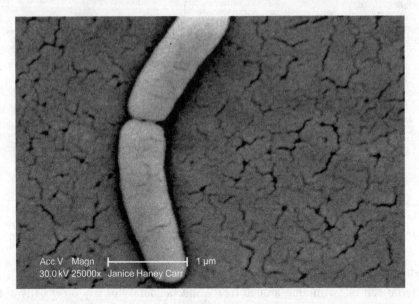

Acc.V   Magn   ⊢⎯⎯⎯⎯⎯⎯⊣   1 µm
30.0 kV 25000x   Janice Haney Carr

Figure 8.1   Bacterial cell division. Colourized electron micrograph of a *Salmonella* bacterium dividing. (Content provider(s): CDC/Bette Jensen; Photo Credit: Janice Haney Carr)

needed before the cell divides. These times are about the same whether the cell is growing quickly on a rich medium, or slowly on a poor medium, even though the time between cell divisions will vary considerably (although, if the growth rate is changed by growing them at a different temperature, that is another story).

When *E. coli* is growing fast, it can divide every 20 minutes. How can it divide more quickly than it can copy its DNA? The answer here is that it starts a further round of replication before it divides so that, by the time of cell division, it is already part of the way through producing the copies of its DNA that will be needed for the next cell division. Thus, cell division does not provide the signal for the start of DNA replication. If we want to get into the question of what the actual signal is, that gets too technical for this book (and indeed is not completely understood), but we do know that it is based on the size of the cell. When the cell gets big enough, it starts another round of copying its DNA, even though the previous round may already be only part way through.

The other side of the question (what triggers cell division?) is rather easier to answer, although again quite complex. This is actually tied in to the other point I posed at the start, i.e. ensuring that both daughter cells have a copy of the newly-replicated chromosome – and, related to that, why does the cell divide in the centre and not asymmetrically? Note that I am thinking here of a rod-shaped bacterium such as *E. coli*.

When the chromosome is copied, this happens – initially at least – in the centre of the cell, and the accumulation of replicating DNA at this position blocks cell division. At a later stage of replication, the two DNA copies get pulled apart towards the ends of the cell, thus freeing up the site where division will occur. That positions the two copies appropriately, but is not enough for division to start. There is another level of control, which was discovered by looking at mutant strains that divide inaccurately, producing some very small cells (*minicells*) that do not contain DNA. These mutants were found to be defective in the genes coding for a set of proteins called Min.

One of these Min proteins is an inhibitor of cell division, while another antagonizes the inhibitor and therefore allows division to occur. The second protein, the antagonist, shuttles rapidly between the two ends, with the consequence that, averaged over time, its concentration is highest in the centre of the cell. At that point, it interacts with the inhibitor protein and stops it from preventing cell division. Thus, division happens at the midpoint of the cell.

So, part of the answer at least is that completion of replication does not provide a positive signal for cell division, but it is necessary for it to happen because it removes a block in cell division, which happens at the midpoint of the cell because of the action of the Min proteins. Also, because the two copies of the chromosome have been pulled to opposite ends of the cell, both daughters will have a copy.

With many bacteria, the simple model is reasonably accurate – the cells grow, copy their DNA and divide, without any clearly defined stages. With eukaryotic cells, ranging from fungi to animal cells, this is not the case. If we consider such a cell growing asexually, including the cell division process called mitosis, there are several well-defined stages, with check points between them (see Figure 8.2).

Following cell division (mitosis, or M phase), the cell enters a phase known as G_1, which can be regarded as a simple growth phase. Progression to the next stage (S phase) is marked by the start of DNA synthesis (chromosome replication). At the end of the S phase, when chromosome copying has finished, there are twice the number of chromosomes and the cell then enters the G_2 phase. This prepares the cell for mitosis, which happens at the end of the G_2 phase; the chromosomes separate and the cell divides.

In rapidly growing human cells, the whole cycle takes about 24 hours but, in a yeast cell, it may take only about 90 minutes. In multicellular organisms, such as ourselves, many of our cells are not growing; they do not enter the S phase, but stay in a state known as G_0, which is a maintenance state. These cells are metabolically active but do not grow or divide, unless specific growth factors are added. Something similar happens in yeast cells, which, in the absence of sufficient food, do not enter the S phase.

This system has a number of checkpoints, which stop the cycle if everything is not in order (as indicated in Figure 8.2). One of the most important things that may go wrong is that the DNA may be damaged, in a variety of ways and by various means – either by external factors (such as radiation) or by mistakes in copying it. If the DNA has been damaged earlier, the cycle is arrested in G_1 phase, which provides a opportunity for the damage to be mended before the DNA is copied.

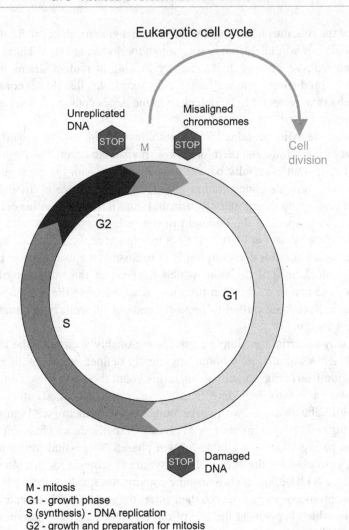

Figure 8.2   Eukaryotic cell cycle M = mitosis (cell division), G_1 = growth phase, S (synthesis) = DNA replication phase, G_2 = growth phase and preparation for mitosis. Stop signs indicate checkpoints.

The knowledge gained from studies of yeast cells, by analyzing mutants that are defective in control of their cell cycle has real practical significance here. In human cells, this checkpoint is mediated by a protein known as p53, and mutations in the p53 gene that prevent G1 arrest if the DNA is damaged are the most frequent genetic alteration associated with the development of cancer. The absence of functional p53, and hence the failure of this checkpoint, allows copying of the damaged DNA. This can lead to the production of cells that have alterations or deletions of key genes that are needed for the control of cell multiplication and differentiation – and hence to the uncontrolled growth of cells that is the basis of a tumour. A further check on the integrity of copying the DNA, and its completeness, is made in G_2 phase, before mitosis, and the accuracy of the separation of the chromosomes is checked before the cell can start G_1 phase, after cell division.

Although bacteria do not have such a defined cell cycle, I do not want to give the impression that all the cells in a culture are exactly the same. In Chapter 1, I described how a bacterial culture goes through several phases (lag, exponential, stationary). However, that is not the complete story. If we have a flask sitting in the incubator (*batch culture*), not stirred or shaken, the conditions within that culture are not uniform. For example, the cells near the surface will get a reasonable supply of oxygen, while those lower down may be virtually anaerobic, so the cells will be responding quite differently. Some may be growing flat out, while others are starved and not growing at all. Also, the conditions within the culture will change over time – oxygen and other nutrients get used up, the pH may change, and so on.

For proper studies of bacterial physiology, we would therefore grow the culture in a fermenter, where it can be stirred thoroughly, and where the supply of oxygen and nutrients can be controlled precisely. If fresh medium is added continuously and an equal volume of spent medium is removed, we have a continuous culture in which the bacteria are kept at a constant level in a constant environment.

Whether we consider batch culture, or continuous culture in a fermenter, the bacteria may not be all identical physiologically. Some will be just about to divide, some will be part way through the cycle, and so on. This may have important consequences. For example, their sensitivity to antibiotics may vary according to their state at any one point in time – in particular, cells that are in the process of division may be more sensitive to antibiotics such as penicillin. We tend to look at the properties of the culture as a whole, which is an average of the properties of all the different individual cells. Sophisticated methods are now being used to enable the study of the behaviour of individual cells in such a culture.

8.2 Motility

Cells such as *E. coli* swim in a liquid medium using flagella. These are long polymers of protein extending from the cell surface. Although they are often described as 'whip-like', they do not act like a whip, but more like a propeller. They are anchored in the cell membrane by an assembly of proteins that acts as a motor and rotates the flagellum, thus driving the cell through the medium. If the flagella rotate anti-clockwise, they coalesce into a bundle and cause the cell to move in one direction in a straight line (a *run*). Periodically, the direction of rotation of the flagella is reversed so that they rotate clockwise. This causes the bundles to fly apart, with the result that the cell tumbles briefly until the direction of rotation is reversed and the cell starts moving again, but now in a different direction. Normally, a run will last about 1 second, during which time the bacterium will move 10–20 times its body length, and a tumble lasts about 0.1 second. The overall effect is of random motion.

In a natural environment, however, the concentration of substrates and other chemicals will vary from one place to another, so it is useful for the cell to be able to move towards a source of food (or away from something damaging). This is known as *chemotaxis*. Specific proteins on the cell surface (sensors) can detect the presence of an attractive chemical, which binds to the sensor. A bacterial cell can respond individually to different chemicals because of the presence of different sensor proteins, each of which will recognize a different chemical. The binding of a chemical to a specific sensor protein causes a change in the structure of the sensor, and this change is recognized by other proteins inside the cell. In this way, these sensors can transmit a signal to the interior of the cell, the consequence of which is an alteration in the frequency of changes in the direction of rotation of the motor. Thus, the length of the runs will be changed, which means that the cell can move towards an attractive stimulus by increasing the duration of the runs, tumbling less frequently so that there is a net movement in that direction.

Much of our knowledge of how this system works, and the nature of the genes and proteins involved, comes from genetic studies such as those described in the previous chapter. It is possible to obtain mutants that are non-motile, some of which are unable to make flagella, while others make partial or non-functional flagella. A further class of mutants make functional flagella, so they are motile, but they do not show specific aspects of chemotaxis. This category includes those that are defective in the production of specific sensor proteins, as well as those that do not properly transmit the signal from the receptor to the flagella. Once suitable mutants have been obtained, we can identify the genes responsible by testing the ability of cloned fragments of DNA to restore motility or chemotaxis to the relevant mutant.

8.3   Biofilms

Previously, I have considered bacteria growing, in a laboratory, on their own. In nature, however, this is the exception; bacteria and other microbes are usually found as part of a complex community. Therefore, as well as considering the behaviour of individual cells, we also need to look at how members of that community interact with one another. An important example of such a community is that known as a biofilm.

A biofilm is a collection of organisms, living as a thin layer on the surface of an object. We are all familiar with biofilms, even if we do not recognize them as such. They range from the plaque that forms on our teeth to the slime that is found on the inside of water pipes and tanks, on the bottom of boats, or on stones immersed in a stream. Less obviously, many of the bacteria in soil exist in biofilms on the surface of particles in the earth. In this situation, the structure of the biofilm is important in retaining a layer of water on the surface of the soil particle. Without this protection, many of these microbes would die in dry conditions.

Let us start by considering plaque on our teeth, which is one of the simplest forms of biofilms. Development of plaque starts with proteins in the saliva sticking to the enamel surface of the teeth, forming a preparatory layer. Saliva also contains large numbers of bacteria, particularly various species of streptococci, and some of these will stick to the protein layer, forming the biofilm that we know as plaque – which therefore consists mainly of a mixture of bacteria, at high density. These bacteria are able to ferment sugars such as sucrose to form acidic products which start to dissolve the enamel of our teeth. Part of the significance of the biofilm here is that the acid can accumulate within the biofilm, since the structure of the biofilm reduces the tendency of the acid to diffuse away from the place where it is formed. Thus, there can be high concentrations of acid locally within the biofilm.

Biofilms can also form elsewhere within the body, especially on bits of plastic that are inserted for various purposes, such as intravenous catheters, artificial heart valves and replacement hip joints. Of course, these are sterile when they are put in, but they can become colonized by the bacteria that often circulate temporarily within the blood, from minor, indeed imperceptible, damage to our gums or skin. Bacteria from the gut also get into the blood in a similar way.

We have very effective defences to deal with these temporary invasions (see Chapter 2) and, generally, the bacteria are cleared within a minute or two. However, if they get a chance to latch onto a foreign body, such as an artificial heart valve, they can colonize it and will then be able to evade the body's defences, which cannot access these sites. One organism that is particularly good at doing this is a species of staphylococcus known as *Staphylococcus epidermidis*. This is a universal inhabitant of the skin surface, which is carried by everybody on their skin. Normally it does no harm at all, and it is much less pathogenic than its closely related cousin *Staph. aureus* – but it really does love to colonize plastic surfaces. It produces a sticky polysaccharide that enables it to do this, thus forming a biofilm.

Staph. epidermidis is particularly a problem with artificial heart valves (or with heart valves that have been damaged in other ways), where the accumulation of bacteria on the surface of the valve can interfere with its action (other bacteria that are important in this context include the streptococci from the mouth, as well as gut bacteria known as enterococci). An additional problem is that bacteria will be shed occasionally from that site, as parts of the biofilm slough off. The body reacts to the presence of these packets of bacteria in the blood, causing intermittent fever.

These biofilms are relatively simple ones, consisting of only a single organism (or a limited range of organisms, in the case of plaque), albeit within a complex structure of proteins and polysaccharides. Environmental biofilms are often much more complex, containing not just a variety of bacteria but other microbes as well, ranging from viruses to protozoa. These organisms interact with one another in a multitude of ways. For example, the waste product from one microbe may be food for another; and those organisms near the surface get preferential access to nutrients (including oxygen) that diffuse in from outside.

The structured organization of the biofilm means that, within the biofilm, diffusion of nutrients, and also movement of microbes and their products, is restricted. Thus, the organisms at the base are in a relatively nutrient poor environment, with limited oxygen, although they will get access to the waste products (including carbon dioxide) from those nearer the surface. This means that the microbes in the lower layers of the biofilm may be growing more slowly (or not at all), and their response to the low levels of nutrient will include changes in the expression of the genes they contain – they may switch off those needed for rapid growth and activate those needed for stationary phase. Alternatively, they may have special 'slow growth' genes, or they may turn on genes needed for anaerobic growth. A further factor is that bacteria can behave differently when there are a lot of them present, causing a range of genes to be switched on or off. This effect, known as *quorum sensing*, is discussed in more detail later on.

The structure of the biofilm therefore leads to the development of different microenvironments within it, so that the predominant microbes may vary from one part of a biofilm to another. In addition, the local accumulation of microbial products, including acids, can have significant consequences. We have already seen, in the case of plaque on teeth, how this effect can lead to the build-up of acid at the tooth surface, and in Chapter 6 we encountered this effect in relation to corrosion.

Biofilms are not only complex communities, but they are also dynamic. As they develop over time, they become thicker by recruiting new cells as well as by the multiplication of those within it. This decreases still further the availability of nutrients, including oxygen, to the cells at the base. Some of these cells may not be able to tolerate such a change, even in a dormant state, so they may die. As it is the cells at the base that are responsible for the adhesion of the biofilm to the underlying material, then if too many of them die, chunks of the biofilm may slough off. Even without this happening, parts of the surface layers of an established biofilm will peel off from time to time. This provides material for the establishment of fresh biofilms on newly available material (such as a stone falling into a stream).

So, biofilms can be an important factor in human infections, as they provide sites that are protected from the normal host defences. Such infections can be very difficult to treat with antibiotics. In part, this can be considered as a failure of the drugs to penetrate into the biofilms, but that is only part of the story. As we have seen, the bacteria themselves change. When they are adapted to living within a biofilm, they may be growing more slowly, or may be dormant. They may switch off some of their genes and switch on others, and these changes in their metabolism cause them to be less susceptible to antibiotics.

Bacteria that are growing slowly, or not at all, tend to be less sensitive to antibiotics. When combined with the inability of the cells of our immune system to access the site, this makes effective treatment exceptionally difficult. The consequence is that, in many cases, the only way of dealing with them is to remove the contaminated bit of plastic. With an intravenous catheter, that is easily done – but an artificial hip joint? It may be necessary for the patient to have to undergo a further hip replacement operation, which can be unpleasant (and potentially hazardous) for the patient and expensive for the health service.

Biofilms are also of considerable practical significance in other ways. They enable bacteria to colonize water distribution systems (pipes, tanks, and so on), and those bacteria include some that can cause human disease (notably *Legionella*, as described in Chapter 3). They are often resistant to normal disinfection processes, such as chlorination (as discussed further below), with the result that the water coming from the taps may contain opportunist bacteria, albeit at a low level. This is unlikely to be a problem in a domestic situation, but can be a serious matter in a hospital, where there are large numbers of highly susceptible people.

A different matter of practical significance concerns biofilms on the bottom of boats and ships. The accumulation of this material, fouling, causes drag that slows the vessel down. In the case of a sailing dinghy used for racing, it is relatively simple to haul it out of the water and give it a good scrub. This is not so easy for a supertanker – and yet the fouling of such a vessel is of considerable economic importance, for obvious reasons. Anti-fouling paints can be used to reduce the problem, but there are environmental concerns, because the antimicrobial content of such paints can leach out into the water and cause damage to the surrounding ecosystems.

There is one important aspect of biofilms that I have not yet considered, which is the presence of protozoa. Protozoa often get their food by eating bacteria, so you have to imagine that this biofilm, a complex structure with various bacteria growing within it, also contains a range of protozoa moving slowly through it grazing on the bacteria. Protozoa ingest the bacteria, kill them and use the nutrients for their own ends. However, as might be expected, some bacteria have developed ways of preventing this from happening.

In the environment, any bacteria that resist attack by protozoa will have an evolutionary advantage, so we would expect this selection pressure to produce species that are able to survive protozoal ingestion. In Chapter 3, I looked at the bacterium that causes Legionnaire's disease – *Legionella pneumophila*. It is widespread in the environment, especially in water, and causes a problem when aerosols are generated, for example from cooling units and showers. It exists within the biofilms that coat the surfaces of pipes and water tanks. The protozoa within the biofilm ingest the *Legionella* cells but do not kill them; the bacteria survive, and multiply, within the protozoal cells.

When food becomes scarce, these protozoa form dormant structures known as cysts, which, being a survival mechanism, have evolved to become resistant to a range of adverse conditions. Unfortunately, this includes disinfectants such as chlorine. Thus, conventional chlorination of the water will not kill the protozoal cysts, nor will it kill the *Legionella* within them – hence the survival of the bacteria within the water systems, and the possibility of outbreaks of Legionnaire's disease if aerosols are generated using the contaminated water. Elimination of the bacteria requires the use of biocides that can kill the protozoal cysts, or draining down the system and scrubbing the tanks to remove the biofilm inside them.

There is a further interesting twist to this story. In Chapter 3, I described how bacteria such as *Legionella*, when they set up an infection, are able to survive and multiply within cells of the immune system such as macrophages. There is an obvious parallel with survival inside protozoa, and detailed studies have shown that many of the genes needed for survival inside protozoa are also necessary for survival inside macrophages. It therefore seems that these bacteria, through a long period of evolution in the environment, have learned the trick of intracellular survival so as to resist protozoal attack, and they have subsequently adapted that mechanism to enable them to survive the attentions of another predatory cell – the macrophage. This story is not unique to *Legionella*. We now know that many pathogens that are able to survive macrophage attack, including those that cause TB, are also able to survive inside protozoa in the environment, thus providing a potential reservoir of infection.

8.4   Quorum sensing

An implication of the above discussion of biofilms is that, in such a community, the behaviour of one bacterium is influenced by that of others around it. This is not always accidental. There are many examples of bacteria 'talking' to one another as a clearly evolved strategy. This is shown most dramatically by situations in which the behaviour of bacteria changes according to how many of them are present – the phenomenon known as *quorum sensing*. It was first observed in some bacteria (*Vibrio* species – non-pathogens related to the organism that causes cholera) that live in the sea, either swimming about freely or inhabiting the light-producing organs of some fish and squids. It is the presence of these bacteria that is responsible for the light produced by those organs – the bacteria are luminescent. However, when the bacteria are free-swimming, they do not produce light.

There are clear evolutionary advantages to this Jekyll and Hyde behaviour. If the bacteria produced light while swimming freely, they would not only be a target for predators but, as it is also an energy-requiring process, they would be wasting scarce resources on an activity that is at best useless and at worst dangerous. While in the light organ of a host, they are protected from predators and they get the food they need from their host to compensate for the energy spent. The host tolerates their presence because the emission of light is useful to them. So there is a good symbiotic relationship.

But how is this change achieved? The answer is that the bacteria only emit light when there are high numbers of them adjacent to one another. This happens in the light organs, where they are packed in quite closely – but when they are free-swimming, the concentration of bacteria is extremely low. This does not fully answer the question; the bacteria must have some way of knowing how many there

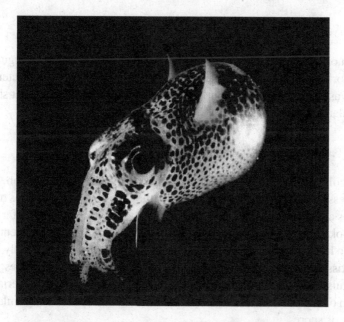

Figure 8.3   A bioluminescent squid (*Euprymna scolopes*) The bioluminescence is due to the presence of a species of *Vibrio* bacteria. (With permission from M McFall-Ngai)

are. This is provided by the secretion of a messenger chemical (technically, an acyl homoserine lactone, or AHL for short), which is produced by each cell. This is secreted into its environment, where it can be recognized by another cell, which responds, if the level of AHL is high enough, by switching on bioluminescence. In the free environment, the amount of this messenger is very low, so there is not enough of it to switch on bioluminescence. However, in the light organ, the large numbers of bacteria present mean that the level of AHL can accumulate, so the level is high enough to activate the bioluminescence genes. Thus, the bacteria can, in effect, count the number of cells present and only respond if there are enough of them around.

There are many other examples of quorum sensing, which has a significant role in situations of practical importance to us. If we consider a pathogenic bacterium infecting a specific site in the body, it makes no sense for the organism to advertise its presence straight away by making an all-out attack; the body would recognize its presence and mount an immune response which would eliminate the invader. It makes much more sense to lie low for a while and allow numbers to build up gradually until a critical mass is reached, *then* start to produce damage.

Quorum sensing controls genes relevant to disease in a wide range of bacteria. These include not only human pathogens, such as *Staph. aureus*, *Pseudomonas aeruginosa* and *Vibrio cholerae*, but also plant pathogens – notably several species of *Erwinia*, which cause soft rot, fireblight and other diseases in a wide range of plants. They make enzymes (pectinase, cellulase) that attack plant tissues, but do so only if there are enough bacteria there to mount a concerted attack.

Also, quorum sensing is by no means limited to pathogens; it is arguably even more important for bacteria in the environment. Its relevance to bacteria within biofilms was mentioned earlier, but it is also essential for microbes that show some multicellular behaviour, and we will come to that later on.

8.5   Bacterial sporulation

On the whole, most bacteria do not show much obvious differentiation. Although gene expression may vary depending on how they are growing, this does not usually show up as any clearly visible differences in a microscope – one cell of a particular species looks very much like another. This is, of course, quite different from the situation in higher organisms, such as ourselves, where there are not only obviously different tissues and organs, but also the cells within those structures can often be distinguished microscopically. But there are examples where bacteria do show some form of differentiation, and one of the clearest examples is in sporulation – the formation of spores.

Species of the rod-shaped bacteria known as *Bacillus* are typically soil inhabitants (although some of them are significant causes of food poisoning – and one, *B. anthracis,* is highly pathogenic, causing anthrax in humans and animals). When nutrients get scarce, these organisms respond by forming dormant spores, which are very resistant to drying, as well as to heat and disinfectants. Another genus that forms resistant spores is *Clostridium*, which contains many significant pathogens including *C. botulinum* (the cause of botulism), *C. tetani* (which causes tetanus), and *C. perfringens* (which causes severe wound infections – gas gangrene – as well as a form of food poisoning).

The process is best understood with *Bacillus*. The spore forms within the cell (and is hence known as an endospore), starting with an unusual form of cell division. Instead of division happening in the centre of the rod-shaped cell, as described earlier in this chapter, a membrane forms towards one end, enclosing a portion of the cytoplasm and a copy of the chromosome (in this case, there is a special mechanism for importing a copy of the chromosome into this small portion of the original cell). So there are now two compartments, each with a chromosome, within the cell – a small pre-spore, and the mother cell, which takes up the rest of the space. A very thick cell wall is produced around the prespore. This is followed by a loss of water from the prespore, which then matures into the final spore structure. The mother cell disintegrates at the end of the process. The spore that has been formed is dormant (it shows essentially no metabolic activity) and highly resistant, largely due to the loss of water. Such a structure may survive virtually indefinitely, and viable *Bacillus* spores have been recovered from the abdominal contents of bees preserved in amber, over 25 million years old.

The whole process is rather more complex than this brief description, and it occurs in a number of stages, but the key point is that all the stages involve different activities happening, in concert, in the mother cell and in the developing spore. As the two compartments go through this process, there is a controlled programme of changes in gene expression. Different groups of genes, in each compartment, are turned off and on, in sequence, and it is essential that the two processes are coordinated. When one stage is completed in the pre-spore, a signal is sent to the mother cell, which switches on a different set of genes, the effect being that the mother cell can contribute, from the outside, to the development of the spore. The mother cell then sends a signal back to the developing spore, which again responds by a further change in gene expression. The molecular basis for these changes involves alterations in the specificity of the RNA polymerase in the two compartments, including activation of different sigma factors (see Chapter 7). Several such cycles result in the final maturation of the spore and the degradation of the mother cell, with release of the mature spore.

In summary, therefore, the development of the spore requires a series of different events in the spore and in the mother cell, and it is essential that the series of events in the two compartments are coordinated so that one stage in the spore does not occur until the mother cell has done its bit, and vice versa. The cross-talking between the two compartments provides an elegant and sophisticated way of ensuring that this is done.

Not all bacterial spores are as heat-stable as those of *Bacillus* and *Clostridium*. In Chapter 6, we encountered the streptomycetes, which are common organisms in the soil – and we will meet them again in Chapter 9, as they are the principal source of naturally occurring antibiotics. These bacteria grow not as single dispersed cells, but as filaments – long, thin structures divided into partially independent cells by cross-walls. In the laboratory, on a solid medium, they will grow in this form for several days, but they then start to produce aerial structures – short branches that stick up from the surface of the medium. At the ends of these branches, chains of spores are formed. These are primarily a dispersal mechanism rather than an aid to survival, and their production at the tips of the aerial branches helps them to be carried to new sites. The change from the lengthening of the filaments to the production of spores is again due, at least in part, to a switch in gene expression caused by the activation of a new sigma factor changing the specificity of the RNA polymerase, so that new genes get switched on.

8.6   Multicellular behaviour

The process of sporulation provides an elegant example of the control of a developmental process, which has provided valuable clues as to how similar processes may be controlled in other organisms. However, it can barely be described as multicellular behaviour. For that we have to turn to other organisms, the first of which are bacteria belonging to the genus *Myxococcus*.

Myxococci are bacteria that live in the soil. Under favourable conditions, when there are plenty of nutrients, they exist as independent cells, swarming around in the films of water that surround particles in the soil. However, when the going gets tough, their behaviour changes. First of all, they come together, forming a mound containing hundreds of thousands of individual cells. Then, within that mound, some of the cells differentiate into spores within a structure known as a fruiting body that is elevated above the surface. These spores are dormant and are resistant to drying; they are thus able to survive in the environment until conditions improve, when they will germinate and form individual cells that resume their swarming behaviour. We thus see a cycle between behaviour as individual cells, when conditions are favourable, and cooperative behaviour leading to spore formation when things are not so good.

It is important for the bacteria to produce spores only when there are enough of them present to form a complete fruiting body, otherwise the process would fail. This is another example of quorum sensing behaviour as described earlier. At an early stage in the process, a signal chemical is made by individual cells, and the process will continue only if the amount of this messenger reaches a critical level. In this way, the bacteria ensure that production of a fruiting body will only happen if enough cells are present.

A similar, but even more dramatic, process is seen with the (eukaryotic) slime mould *Dictyostelium*. This also exists in the soil as a single-celled organism, feeding saprophytically on rotting vegetation, but it switches to a multicellular organization when food becomes scarce (or other conditions become less suitable). As with *Myxococcus*, the cells aggregate, in this case using a chemical signal they secrete so that they can find one another. They can sense the concentration of this chemical and move up the concentration gradient so that they move towards the position where the level is highest – which corresponds to a site with a number of cells close together. The aggregated cells form a coherent mound, which then develops into a structure that resembles a tiny slug (a few mm long); this is capable of moving around, thus behaving as a genuinely multicellular organism. Eventually, it settles down and the cells within it differentiate to form a base plate, a stalk and a fruiting body. The spores within the fruiting body are able to survive adverse conditions, and they will germinate eventually to produce single-celled organisms again.

This system has been of great interest to those studying developmental processes in higher organisms, as it provides a simple model system for analyzing how cells become committed to forming different parts of the final structure. All of the cells are initially identical, so how do some end up forming the stalk and some form the fruiting body? At a simple level, the answer is that the fate of a cell is determined by its position and its movement in the early stages of transition from the slug to the final structure. Chemical signals secreted by the cells influence the development of those around them. This is, in some senses, analogous to the differentiation of cells during human development.

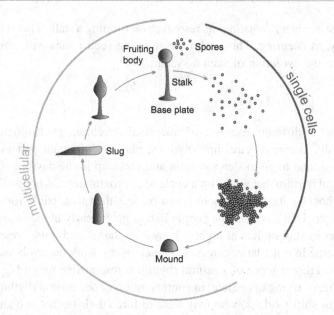

Figure 8.4 Life cycle of the slime mould *Dictyostelium* Redrawn after Gaudet *et al.* (2008), BMC Genomics (**9**), 130. Published online. doi: 10.1186/1471-2164-9-130.

A further interesting feature is that there is an optimum size to the stalk. If it is too short, the spores are not so readily dispersed – but the longer it is, the fewer cells remain to form spores, and therefore fewer cells will survive. Although the cells are committed at an earlier stage to becoming either stalk or spore, there are further checks made on this ratio, as can be shown by removing some of the precursor stalk cells. The response will be that some cells which were originally headed toward becoming spores will now change track and become stalk cells instead, thus maintaining the optimum ratio of stalk to spore.

Dictyostelium is also of interest to those studying evolution. Of the initial aggregate of cells, only those that end up as spores will survive. We are used to thinking of evolution, and the 'survival of the fittest', in terms of whole organisms, but there are complications when we consider the apparently altruistic behaviour of some individuals in a population (similar considerations apply if we think about, for example, social insects such as ants and some bees and wasps, where only the queen lays eggs, so many of the individuals leave no progeny behind).

There is one aspect of this that is especially interesting. It is possible, in the laboratory, to produce mutants of *Dictyostelium* that are unable to form stalks. Such a mutant would not propagate well in the environment, because its fruiting body is not raised up. On the other hand, since all the cells become spores, it forms a lot of spores. If wild type and mutant strains are mixed, they can produce a chimaeric structure, in which the stalk cells are all wild type but the fruiting body contains a mixture. In that mixture of spores, there are more mutant cells than wild type, since the mutant strain is not 'wasting' its resources on forming a stalk. This is known, not surprisingly, as *cheating*. Those who are studying such issues will argue at great length about the evolution of such a system.

8.7 Biological clocks

This is a rather different example of microbial development. Biological clocks, more formally known as circadian rhythms, play an important part in our lives. Basically, we tend to go to sleep at night and wake up in the daytime. Our bodies have a natural rhythm that works on a cycle of approximately 24 hours. But it is not exactly 24 hours – for some people it is a bit less than that, while for others it is slightly longer. That is why some people like to get up early in the morning, while others prefer to stay up late at night – known as 'larks' and 'owls' respectively. I definitely come into the latter category – much of my work on this book was done late at night. The existence of a natural rhythm is responsible for jet lag, where we are operating, or trying to operate, in contravention of our natural rhythm. It is also important for shift workers, who may have to force their bodies into an unnatural rhythm.

You might think that if someone's clock operates on a cycle that is different from 24 hours, they would gradually get more and more out of phase with the cycle of day and night, but this does not usually happen, due to a phenomenon known as *entrainment*. Put simply, this means that we respond to the presence of light by re-setting our biological clock, so that it keeps in time with the external cycle. When we travel to a different time zone, the difference may be too great for this to happen straight away, so it may take a few days to get used to the new time.

Why is this relevant to bacteria? After all, most of the bacteria that we study in the lab have a generation time that is much shorter than 24 hours. *E. coli*, for example, would typically divide every 20–30 minutes, so it would go through very many generations in a whole day. Surprisingly, though, some bacteria do have a circadian rhythm despite this. Part of the reason is that a bacterium does not die when it divides; it just makes two cells out of one, so that each of the daughter cells still contains part of the cytoplasm of the parent cell. Thus, a bacterial colony, or a collection of cells in a liquid culture, can still be regarded as a single 'organism' for this argument. In other words, we can regard a microbial culture, even if growing as separate cells, as a multicellular organism in which the individual cells are moving around freely rather than being organized into a single 'body'.

Some of the best-studied examples of biological clocks in microbes come from the photosynthetic cyanobacteria. The advantage of responding to the presence of light is obvious for a photosynthetic organism, but it goes further than that. These bacteria show all the typical features of a genuine biological clock, in that their response goes up and down even under constant light conditions. As with our own clocks, the role of the light/dark cycle is to entrain the natural cycle and keep the clock on a constant 24 hour cycle. Using mutants with altered rhythms, a group of three genes known as *kai* (from the Japanese word for rotation or cycle) has been shown to be essential for the operation of the clock. Genes with similar sequences have been found in many bacteria and Archaea, suggesting that such biological clocks are a widespread phenomenon in many other organisms that have not yet been studied in detail.

This chapter and the previous one have been concerned largely with fundamental biological questions. In the next chapter, I want to return to practical matters, albeit drawing on some of this knowledge, which is the practical use that we make of microbes (other than in food) – that is, microbial biotechnology.

9
Microbial Biotechnology – Practical Uses of Microbes

Biotechnology has become a controversial word. The reasons cited for opposition to biotechnology, and especially to genetic modification, are many and varied – but they are mostly without any scientific justification. In this chapter, I will describe some of the multitude of ways in which microbes have been, and continue to be, used for practical purposes.

The major use of microbes, by a long way – both in the extent of their use and the time over which they have been used – is in food preparation. This includes bread, fermented drinks (beer, wine, cider, etc), cheese, yogurt and many others. These were described in Chapter 5, so I don't need to go over them again – except to remember, in the context of considering the impact of biotechnology, that we have been happily consuming such products for thousands of years and continue to do so. Indeed, in some cases, we consume substantial numbers of microbes along with the food ('friendly bacteria'), which is advertised as being advantageous to health.

More recently, microbes have been used to make specific products, notably antibiotics. Although such processes originated with natural microbes, in general the level of product that they make is much too low to be practically useful, so it was necessary to select strains with higher yields. This included using various mutagenic agents such as UV radiation to cause random mutations in the chromosome, followed by testing the resulting mutants for variants with higher yields. Subsequently, with the advent of molecular biology techniques (as introduced in Chapter 7) and with greater knowledge of the genetics and biochemistry of these organisms, it became possible to produce specific mutations rather than random ones.

This is the first major difference between what we now call 'genetic modification' and the methods previously used to obtain genetic variants. With modern genetic modification, we know precisely what changes have happened, whereas, with the

Understanding Microbes: An Introduction to a Small World, First Edition. Jeremy W. Dale.
© 2013 John Wiley & Sons, Ltd. Published 2013 by John Wiley & Sons, Ltd.

older methods, the changes were random and unknown and would have been likely to have included a number of changes other than what we wanted. Correspondingly, despite the controversy over the safety of genetically manipulated organisms, there is much more uncertainty attached to conventional approaches.

The second major difference arising from the new technology is that it has become possible to add new genes. With traditional techniques, it has only been possible to increase the amount of product formed, or sometimes to get minor changes to it. With the new methods, we can get bacterial strains to make a product that is totally different from anything that they would make naturally. This includes, especially, getting microbes to produce proteins that are naturally made by humans and other animals.

I will look at some examples of both types of process, but it is worth remembering that genetic modification is, in some senses, not new. Virtually all of the food we eat is derived from animals or plants that have been subjected to genetic alteration over tens of thousands of years. One has only to look at a cow, or an ear of wheat, to realize that it is markedly different from its wild ancestors.

9.1   Amino acids

Amino acids, which are the constituents of proteins, are one of the most important of microbial products, at least in terms of volume. Lysine is made in large quantities (over 30,000 tons annually) as a supplement for cereal-based animal feeds which are deficient in this essential amino acid (an essential amino acid is one that an animal cannot make for itself but which has to be provided in its food). The bacterium used, *Corynebacterium glutamicum*, normally produces only limited amounts of lysine, as the enzymes responsible for making it are shut off in response to the accumulation of lysine – this is known as *feedback repression*. The organism also uses intermediates in the pathway for making other amino acids, which are not wanted in this process. However, it is not difficult to produce mutant strains that do not have feedback regulation and also lack the enzymes needed for making the unwanted by-products. Such a strain will make large amounts of lysine – over 50 grams per litre – which sounds a long way short of 30,000 tons until we factor in the enormous size of the fermenters used for growing these strains.

Another commercially important amino acid, produced on an even larger scale (>200,000 tons annually), is glutamic acid. The most familiar use of this is as the flavour enhancer monosodium glutamate. Glutamic acid is also made by a strain of *C. glutamicum*. One of the limitations in the natural process in this case is that intermediates in the pathway are used by the cell for other purposes. Mutants that lack an enzyme responsible for this unwanted activity can concentrate their effort in the production of glutamic acid.

There are many examples of products other than amino acids, such as citric acid, used as a preservative and flavouring (250,000 tons per year). This is almost

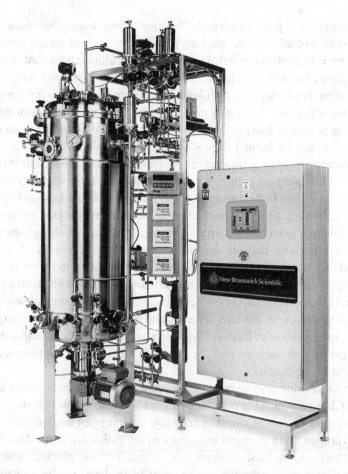

Figure 9.1   A commercially available fermenter for growing microbial cultures, up to 3,000 litres. (Courtesy of Eppendorf AG. All rights reserved. Copyright © 2012.)

all produced microbiologically, by the mould *Aspergillus*. The versatility of microbes, and the ease of growing them in large quantities, makes them a very attractive resource for large scale production of food additives, as well as for other uses.

9.2   Biofuels

There are many ways in which useful fuels (such as ethanol, methane, or hydrogen) can be obtained from biological materials, including agricultural products and plant waste. As this book is about microbes, I will limit the discussion to processes in which microbes (or microbial enzymes) are involved.

The simplest process uses fermentation by yeasts to produce ethanol, with, for example, sugar from sugar cane as the starting material. This is essentially the same process as used in brewing or wine making (see Chapter 5), and it is capable of producing up to 12 per cent ethanol. There are serious limitations to the use of this process, not least of which is the relatively low concentration of ethanol produced. The problem here is that ethanol is lethal to the microbes; although tolerant strains can be produced that can make higher levels, it is still far too low to be used directly as a fuel. The ethanol can be extracted by distillation, but this requires heat – fine for high value products like whisky, but for energy production it is wasteful. It can even result in less energy being produced than goes into the process, which is obviously pointless unless waste material can be burned to produce the heat needed. Another, less obvious, limitation of using yeast fermentations is that they cannot use starch (the major component of maize) or cellulose (which forms the bulk of many other plant products). These have to be broken down to usable sugars by heat and/or enzymes.

A quite different process involves anaerobic digestion (i.e. breaking down material, ranging from farm slurry to domestic waste, in the absence of air). This involves bacteria that are unable to grow if air is present (they are strict anaerobes) and which produce methane as a result of this activity (or more strictly biogas, which contains 50–75 per cent methane, together with hydrogen, carbon dioxide and other gases). This is similar to the process of digestion of sewage sludge (which will be dealt with in a later section), and it can be carried out in purpose-built tanks (e.g. for digestion of farm slurry) or in specially designed landfill sites which are lined (to prevent seepage into the ground) and capped so as to enable the gas produced to be captured.

Some bacteria, and also some algae (especially photosynthetic ones) are able to produce hydrogen, which is a potentially useful fuel and a very clean one, as burning hydrogen produces only water. Alternatively, the enzymes needed for the production of hydrogen from biomass material can be isolated and used in a cell-free process which is, at least theoretically, a much more efficient process. One limitation is that most enzymes are inactivated by heat, so the temperature has to be kept relatively low (30–37 °C). However, using enzymes extracted from thermophilic microbes, which can grow at temperatures up to 120 °C (see Chapter 6), has the potential to circumvent that problem. Enzymes from such microbes are more heat-stable, which would enable the process to be run at a higher temperature, thus speeding up the reaction rates dramatically.

In the longer term, microbial fuel cells offer an interesting prospect. A fuel cell is like a conventional battery, in which electrons are produced by a chemical reaction at one terminus (the anode) and flow through an external circuit to the other terminus (the cathode), where a second reaction soaks up the electrons, A fuel cell differs from a battery in that the material required for the reaction is supplied continuously, as required, so it doesn't run out like a battery does.

The advantage of fuel cells is that they are potentially much more efficient in generating power than can be achieved by simply burning the fuel, either in an internal combustion engine or by using the heat to generate steam to drive a turbine. In chemical fuel cells, the reaction at the anode may be by oxidation of hydrogen to water (plus electrons) while, at the cathode, oxygen is reduced by the addition of electrons. In a chemical fuel cell, these reactions are driven by the use of suitable catalysts, whereas, in a microbial fuel cell, microbes (principally bacteria known as *Geobacter*) undertake the oxidation of hydrogen (or a variety of other materials), generating electrons, which are passed on to the anode, thus producing the current. The anode, which may be made of a solid metal oxide, thus acts as an electron acceptor – or, in other words, it is reduced to the metal by the reaction.

For the fuel cell to work, the current has to flow through an external circuit back to the other electrode (the cathode). At this end of the cell, the requirement is for the microbes to oxidize the electrode (remove electrons from it) and pass those electrons on to a suitable substrate such as oxygen or hydrogen ions – in the latter case, converting the hydrogen ions to gaseous hydrogen. Nitrate can also be used as an electron acceptor, which suggests another possible use for such a cell, namely the removal of nitrate from waste water.

9.3 Microbes and metals

Microbes are important in the natural cycling of metals in the environment, both in liberating metals from minerals in rocks and in the reverse process – the deposition of minerals. The role of microbes in the corrosion of metals (Chapter 6) is part of this cycling of metals, and in the same chapter I also described the ability of some microbes to flourish in the toxic environments of mine spoil heaps. Here, I want first to look at how these processes can be exploited, firstly in the recovery of metals from low-grade materials, such as poor quality ores or old mine wastes (bioleaching, or biomining). Later on, I will consider the removal of contaminating metals from soil or water.

Metals in rocks are mostly found as ores, in which the metal is chemically linked to other elements. The simplest form, which will serve as an example, is a combination with sulphur, such as iron pyrite (FeS_2). Mostly, these ores are hard and insoluble minerals, so the job of the microbe is to extract the metal and make it soluble. In the case of iron pyrite, this can be done by oxidizing the sulphur to sulphate. A wide variety of bacteria can be used for this purpose, so I will just single out one of them, which rejoices in the name of *Acidithiobacillus ferrooxidans*. This not only oxidizes the sulphur in iron pyrite to sulphate, but will also oxidize the iron from its ferrous to ferric form. This is relevant, because the real merit of the process lies in extracting more valuable metals such as copper, and the ferric ions will in turn oxidize the copper sulphide, thus making it soluble. A further essential feature of this bacterium, and other bacteria used in this way, is that they are able to survive and grow in the highly acidic conditions that are created.

A substantial proportion (some 25 per cent) of the worldwide production of copper now involves microbiological treatment of either low-grade ores or the waste left behind after the better quality ore has been used. One common way of doing this is to gather the material to be treated into a large mound, on an impervious surface, and spray it with the bacteria in an acidic solution. The dissolved copper is

Figure 9.2   Mineral contamination. Mineral Creek, Colorado, USA, receives acid mine drainage from abandoned mine lands. The pH of the stream was about 3.0, and the streambed was heavily coated with aluminium and iron precipitates. (Credit: U.S. Geological Survey Department of the Interior/USGS U.S. Geological Survey)

recovered from the liquid at the base of the heap, and the rest of the liquid (now highly acidic and containing the bacteria and the ferric ions) is returned to the top of the heap. The extraction of metals from heaps is a relatively simple and inexpensive process, but it does suffer from the disadvantage that it is not easy to control the conditions within the heap – especially the temperature. Nor is it possible to control the microbial content; indeed, the earliest applications simply used naturally occurring microbes. Greater control can be achieved by carrying out the reactions in tanks (bioreactors), but this is more expensive and only limited amounts of material can be treated at a time.

Microbes also play a major role in cleaning up the mess left behind by various industrial processes. This includes the waste from mines. In the processes described above, microbes are used to recover metals from such waste, under controlled conditions, so that the resulting liquid can be collected. However, similar processes will happen naturally in the random piles of debris left around the mine, resulting in run-off water that is both highly acidic and also rich in potentially toxic metals. Thus, microbes are part of the problem, but they can also be part of the solution. In this case, sulphate-reducing bacteria can be used to eliminate the metal ions (by making them insoluble – the reverse of the process described above). This can be combined with a reed-bed system that reduces the acidity of the run-off, as well as filtering out the insoluble metal-containing particles.

Metal ion contamination is also a problem in connection with the decommissioning of smelters and other metal-processing plants. I grew up in Swansea, in South Wales, where there was a long history of metal smelting. By the 1960s, almost all the factories had closed, leaving a landscape of derelict buildings and barren wasteland. Nothing would grow. The soil was too toxic, with an accumulation of, for example, arsenic and lead, as well as being highly acidic and not retaining water because of the slag. Some of the neighbouring houses had made a brave attempt at a garden, with the pathetic result of bare earth containing a few sickly plants. However, the area has now been transformed into a pleasant and productive landscape, and this has been achieved partly by biological detoxification, using both microbes and plants that were found to be able to tolerate the highly toxic conditions.

Not all contaminated sites are as bad as that but, even so, the problem has to be dealt with if the site is to be used for other purposes, especially building houses. Not only is removing the soil and dumping it somewhere else expensive, this also does not really solve the problem. It just transfers it elsewhere.

A major factor is the mobility of the pollution. If it is freely soluble, it will probably have been washed out before we even start. Conversely, if it is not soluble at all (or at least not likely to be released within a reasonable timescale), it is again not really a problem; it is not really any different from the rocks all around us, many of which contain potentially toxic minerals locked up in an insoluble form. The problems arise from contamination that is slowly released. This gives us a clue as to the choice of strategy for dealing with the contamination. The soil could be treated to release the contamination, which might involve scooping it up and reacting it with microbes in large tanks or ponds, before returning it. Or, alternatively, it could be treated *in situ* in such a way as to immobilize the metals permanently, by converting them into an insoluble form.

As a further example, we can consider the remediation of the 2012 Olympic Games site in East London. This land had been used for over 150 years by a variety of industries, including several (e.g. car breakers and chemical plants) that generated a wide range of contaminants, such as petroleum products and other hydrocarbons, heavy metals (e.g. lead), arsenic and cyanide. Additional problems included the use of parts of the site for landfill dating back to the Industrial Revolution, plus demolition material from buildings in London that were damaged in World War II, and rubbish tips from the 19th and 20th centuries. About two million cubic metres of soil was excavated; most of this was treated on site, and re-used.

If you have difficulty visualizing two million cubic metres, think of an area 1 km × 2 km, which is 200 hectares, and dig it out to a depth of one metre. Of this, about 30,000 m^3 was treated by bioremediation, using a specially constructed 'bed' in which air was introduced at the base and the temperature controlled by circulating hot and cold water. Although this was only a small proportion of the total amount treated (the remainder being treated by washing or in a variety of other ways), it still represents a major application of the use of microbes for remediation of contaminated soil.

9.4   Oil spills

On 20 April 2010, an explosion wrecked the Deepwater Horizon oil rig in the Gulf of Mexico. The rupture of the pipe from the well caused extensive release of oil into the sea, and it is estimated that, by the time the leak was plugged on 15 July, some 780 million litres (4.9 million barrels, or about 900,000 tonnes) had been spilt. This caused fears of extensive and long-term damage to the environment, especially the fragile coastal wetlands of Louisiana, and to the fishing and tourist industries of the region. How real are those fears? The short-term effects are undeniable and very visible, especially oil-covered sea birds and contaminated shorelines. But to predict the long-term effects, we need to think about what happens to the oil.

Crude oil is a mixture of many components. Some of these are light and volatile, and will escape rapidly by evaporation when they reach the surface of the water. It is the heavier components that are potentially more damaging. Can these be broken down by microbial action? One fact that is often lost sight of is that oil is a natural product, formed over millions of years by the pressure of sediments laid down over organic material formed by the decomposition of plants and animals – and microbes as well. The reservoirs that we tap into to recover the oil only happen because the layers above form a dome of impervious rock. In other circumstances, the oil will seep out into the sea (or elsewhere). It is estimated that about 140,000 tonnes of oil seeps naturally into the Gulf of Mexico each year, but of course this is widely distributed so it does not have the same sort of effect as an oil spill. However, it does mean that microbes are naturally exposed to this material and, over millions of years, some have evolved to be able to take advantage of what represents a valuable food resource for those that are able to use it. Thus, there is no shortage of bacteria and other microbes that can degrade crude oil.

Has this actually happened? One piece of evidence following the Deepwater Horizon incident comes from claims to have detected underwater regions with abnormally low oxygen content, which would be consistent with extensive microbial action degrading the oil (and using up the available oxygen to do it). An official estimate in early August 2010 (less than four months after the incident) claimed that the vast majority of the oil had already disappeared – due not only to natural degradation but also the effects of skimming, burning and direct recovery. However, these claims are still controversial, and there may be extensive underwater plumes of oil, together with deposits in sediments where degradation is low.

We can try to get some sort of pointer to the future by examining previous incidents. In June 1979, a very similar fate happened to the drilling rig Ixtoc I, operating off the coast of Mexico – there was a well blow-out, the blowout preventer failed, the rig caught fire and sank, and oil gushed out of the damaged riser. By the time the well was capped, in March 1980, 475,000 tonnes of oil had been released. There was a devastating short-term effect on local fishing but, within a few years, catches had returned to normal. Sandy beaches and rocky shores recovered quickly (although patches of tar remained above the tide line), but the mangrove swamps were more badly affected. Oyster beds have still not recovered.

Although the Ixtoc I incident closely parallels that of Deepwater Horizon, there are important differences. Ixtoc I was operating in 50 metres of water, while Deepwater Horizon was drilling 1.5 km below the surface. Also, the Louisiana wetlands are a highly vulnerable environment. So we cannot compare the two incidents too closely. Time will tell.

Oil tanker accidents are also instructive. In February 1996, the *Sea Empress* grounded in the approach to Milford Haven, in Wales, releasing about 72,000 tonnes of crude oil, close to important nature reserves especially seabird colonies. About 100 km of coastline was seriously polluted and thousands of sea birds died. Within 12 months, most of the pollution had disappeared, aided by the rocky shoreline, the light nature of the oil spilt and an intensive clean-up campaign. On the other hand, when the *Exxon Valdez* hit the rocks off the coast of Alaska, in 1989, releasing 37,000 tonnes, the pollution was much more persistent, partly because the cold water there slowed down the microbial degradation, and partly because the oil was not dispersed by wave action so effectively, so it aggregated into lumps that were less easily broken down by bacteria.

The general message is that, under suitable conditions, 50 per cent or more of the oil released may be degraded by microbial action. Is it possible to help the process along? The obvious approach would be to spray the oil slick with suspensions of oil-degrading bacteria. Unfortunately, though, when this has been tried, the results have been rather disappointing. This is probably because the naturally-occurring microbes are already present and able to do the job much better than the organisms we produce in the lab. One factor that limits microbial degradation of oil is that the bacteria require other nutrients in order to grow. The oil provides plenty of carbon and energy for growth, but bacteria need other things as well, especially a source of nitrogen and phosphorus. So a possible answer is not to add bacteria, but to use fertilizers to enhance the activity of the microbes that are already there.

The other factor is that oil and water don't mix – and the bacteria are in the water. Therefore, they can only attack the oil at the interface between the two. If the oil is well dispersed – either naturally, for example by the action of waves on a rocky shore, or by the use of chemical dispersants (as long as they are not themselves toxic) – then the microbes can get to work. If it aggregates into large lumps, there is much less scope for the microbes to do their job.

9.5   Sewage and water treatment

Now let's turn to another potential source of pollution – the waste water that we generate in our homes. We empty the bath, or flush the toilet, and forget about it. If this were to be just discharged into the rivers then, apart from being smelly and unsightly, it would create severe environmental problems due to what is known as the oxygen demand. Domestic sewage contains both suspended solids and dissolved organic matter, and it provides a rich environment for microbial growth. If it were to be discharged into rivers, the massive growth of bacteria and other microbes would rapidly consume the oxygen in the water, so fish and other life in the river would die. The first aim of sewage treatment is, therefore, to reduce the oxygen demand so that the effluent can be discharged safely. Microbes are central to this process.

Before the microbes get to work on it, the waste water is passed through a metal grille to remove large objects, and then it flows through channels to allow grit to settle out (the domestic waste is, by this time, mixed with surface water run off from roofs and roads). A settling tank allows some of the suspended solids to be physically removed, and then the microbial action can start.

There are several quite different processes that can be used. One of the most familiar, if now rather old-fashioned, is called a trickling filter. This consists of a circular tank filled with stones or clinker, onto which the settled sewage is sprayed

Figure 9.3 Sewage treatment. A trickling filter at the sewage-treatment plant, Henderson, North Carolina, USA. (Photograph by Brian Hayes from Infrastructure: A Field Guide to the Industrial Landscape, W. W. Norton, 2005)

by means of rotating arms. Despite its name, this system is not really designed to filter the sewage. As the liquid penetrates downwards through the bed of stones, a complex biofilm develops spontaneously on the surface of the stones, containing a mixture of bacteria, fungi, protozoa, algae and even larger creatures such as worms. These feed on the rich mixture of nutrients in the sewage and convert it into biomass. The open structure of the bed, and the slow trickling of liquid downwards, allows air (admitted at the base of the tank) to pass upwards, so the whole process remains aerobic. The liquid that emerges at the base has a much reduced oxygen demand and may be fit for discharge without further treatment. As the biofilm develops, layers will become detached, but these aggregates will settle out quite easily in a settling tank before the liquid is discharged.

A more modern alternative is known as the activated sludge process. Here, a tank is filled with settled sewage, and some sludge from a previous treatment is added to provide the microbes needed to digest the sewage. The tank is stirred continuously and/or air is bubbled through it to make sure that the microbes have enough oxygen to complete the job. As with the trickling filter, the microbes convert the nutrients into biomass (so the number of microbes increases substantially) and, at the end of the process, the microbes (which have now aggregated into lumps known as flocs) are allowed to settle out before the liquid is discharged. The settled material forms the sludge, part of which is returned to the tank along with the next batch of sewage. The process automatically selects the most effective microbes. Those that grow best, and those that settle out most effectively, will be over-represented in the sludge that is returned to start the next batch.

An activated sludge process can also be run continuously, by introducing the sewage, and the sludge that is used to seed it, at one end of the tank and allowing it to flow along the tank. The flow rate is adjusted to ensure that, by the time the material reaches the far end of the tank, the microbes have finished their work, the nutrients have been used up and the liquid is in a fit state for discharge. In this system, known as a plug-flow bioreactor, the tank is designed and operated in such a way as to minimize any horizontal mixing of the contents.

So, with either system, we have converted the initial sewage into a clear, colourless, odourless liquid, with a low oxygen demand, which is fit for discharge into a river or into the sea. I remember visiting a sewage plant and the guide proudly showing us a glass containing some of the effluent, which he said was fit to drink. However, he didn't take up the challenge to prove the point!

However, this is not the end of the story. The microbes that have done the job are now contained in the sludge that has settled out, and something has to be done with them. The UK produces about a million tonnes of sludge per year. The old-fashioned approach was to load it into barges and dump it out at sea – not a good idea and, indeed, one that is no longer allowed in the EU. Landfill and incineration also pose considerable problems. One of the best methods is to subject it to anaerobic digestion – that is, incubating it in a tank in the absence of air. Under these conditions, a different set of microbes flourish and produce methane, which can be used to produce much or all of the power needed to run the plant. There is still a sludge produced, but the volume is now much less.

Two problems have not been considered. The first is, what happens to any pathogenic bacteria or viruses that may be present in the initial sewage? The process is not specifically designed to deal with pathogens, being more concerned with factors such as nutrient levels and oxygen demand, as well as subjective ones such as appearance and smell. The numbers of pathogens are likely to be substantially reduced, but complete elimination of bacterial pathogens is not guaranteed, and viruses are even more problematic. It is therefore a good idea to make the effluent safe by chlorination or by ultraviolet irradiation.

The second problem is chemical contamination of the sewage. At the domestic level, this would include flushing unwanted drugs and chemicals down the toilet (which you shouldn't do, but it does happen). Also, what happens to any medication being taken, including antibiotics as well as oral contraceptives and other hormones? Some of it gets broken down in the body, but a substantial proportion is excreted. What do all these chemicals do to the microbes that run the sewage plant? Fortunately, at the domestic level, they get diluted so much by all the other material that gets into the sewage that they do not really affect the operation of the plant. However, there is some concern over the effect that these chemicals, especially hormones, have on the life in the river into which the effluent is discharged. Additional controls are needed to reduce the risk from large users such as commercial and industrial premises – including university research labs, where it is not unknown for PhD students to dispose of surplus reagents down the sink! And that still leaves possible uncontrolled contamination from run-off water from the roads.

One aspect that does need special mention is contamination by heavy metals – lead, copper, zinc and so on. These can be accumulated by the microbes in the sewage plant and thus find their way into the sludge. If the intention is to dispose of the sludge by using it as a soil conditioner on agricultural land – which seems a good, eco-friendly way of getting rid of it – it is important to make sure that it does not contain unacceptable levels of metal ions or other toxic material.

9.6   Antibiotics and other medical products

I now want to return to the use of microbes to produce useful products. The serendipitous discovery of penicillin in 1928 by Alexander Fleming, who observed inhibition of staphylococcal growth around a contaminating fungal colony on an agar plate, is well known and was described in Chapter 4. Fleming was not the first to observe such a phenomenon and, like previous observations, his would have remained as a curious observation if it had not been for the (less well known) work of Howard Florey and Ernst Chain, who purified the penicillin and characterized its effect, especially demonstrating that it was non-toxic and was able to cure infections. In collaboration with a team of researchers, they also developed methods for mass production of this invaluable antibiotic, which saved many lives during the Second World War.

However, the amounts available were still very limited, as the producing organism (in this case the common mould *Penicillium*) only made very low levels of penicillin. Producing enough for it to be made available to the general population required an extensive strain improvement programme. Let's put some figures on this to illustrate the problem. The first step was the isolation, in 1943, of a different strain of *Penicillium* that naturally produced higher levels of the antibiotic – about 60 mg per litre. This was a considerable improvement on Fleming's original strain, but it is still not a lot. One day's treatment requires at least 600 mg of penicillin, so they needed ten litres of culture to make enough penicillin to treat one patient for one day. To improve the yields, they looked first for spontaneous variants with higher yield – which doubled the amount it was possible to get – and then used various mutagenic agents to produce more variants. Within a few years, they had managed to increase the yield tenfold, to some 600 mg per litre. Many further substantial increases in strain capability have occurred since, combined with improvements in the methods for growing the producing strain to maximize yield, so that modern penicillin production is believed to yield 20–40 g per litre of culture.

The success of penicillin stimulated a search for other natural products with antimicrobial action, by screening soil samples from all over the world for organisms that made antibiotics. The first success came in the late 1940s, with the discovery of streptomycin by Selman Waksman. Streptomycin was a very useful addition to the armoury, as it has a broader spectrum of activity than penicillin – that is, it is able to kill some bacteria that are not susceptible to penicillin. Notably, it was the first effective treatment for tuberculosis.

Streptomycin is produced not by a fungus but by a filamentous bacterium, *Streptomyces*, that lives in the soil. We encountered this group of bacteria in Chapter 6. They are amazingly prolific when it comes to producing antibiotics (and other potentially useful compounds), and most of the subsequent discoveries, such as chloramphenicol (1947) and tetracycline (1948) also came from *Streptomyces*, using the same empirical screening procedure – that is, simply testing samples of soils for organisms with the ability to kill or inhibit bacteria. However, chloramphenicol is now made exclusively by chemical synthesis, rather than being extracted from *Streptomyces* cultures.

Further successes followed, but this was then complemented by the ability to modify the natural products. For example, once the structure of penicillin was known, it was possible to use an enzyme to remove part of the molecule (known as the side chain, because it sticks out from the main part of the structure), which could then be replaced by alternative side chains, thus making a whole series of new semi-synthetic penicillins. One of these, methicillin, was valuable because of its activity against *Staph. aureus* strains that were resistant to the native penicillin (see Chapter 4). Another early example was ampicillin, which extended the range of bacteria against which this group of antibiotics be used. There is now a whole family of antibiotics related to penicillin.

The hunt for new antibacterial antibiotics has slowed down in recent years, apart from targeted programmes aimed at specific organisms, but empirical screening of environmental isolates for useful biological activity continues with a wider brief, to include a range of potential targets. Examples include anti-cancer drugs, where the anthracyclines (originating from *Streptomyces* bacteria) are amongst the most effective drugs. Other examples include statins, widely used to combat high cholesterol levels (which originated from various fungi or bacteria, although many are now produced synthetically) and immunosuppressants (used, for example, to prevent rejection of transplants). One of the earliest immunosuppressants, cyclosporin, was found in cultures of a mould.

Later in this chapter, we will see how microbes can be used to make an even wider range of therapeutically useful products, using gene cloning technology.

9.7   Vaccines

Vaccines have come a long way since their experimental use by Edward Jenner in 1796 (see Chapter 4). We now have safe and effective vaccines against a wide range of viruses, and also some bacteria. In Chapter 4, I looked at the use of vaccines; here I want to deal further with the ways in which vaccines are developed and tested.

Jenner's smallpox vaccine was a live, attenuated, vaccine – that is, although it was a live virus, it was incapable of causing smallpox. However, it was able to provoke an immune response and, hence, protect the subject against the disease. Jenner was fortunate in being able to identify a naturally occurring attenuated virus; more commonly it is necessary to produce an attenuated strain of the pathogen in the laboratory.

Some pathogens will become attenuated spontaneously, if subcultured repeatedly in the laboratory. For example, the TB vaccine BCG (see Chapter 3), Bacille Calmette-Guerin, was developed in this way, named after the two French scientists who developed it at the Institut Pasteur in Lille, Albert Calmette and Camille Guerin. The procedure they used to develop the vaccine was based on the idea that growing the bacterium on a medium to which it was not well suited would eventually reduce its virulence. The bizarre medium they chose consisted of slices of potato soaked in ox bile. They subjected the TB bacterium to subculture on this medium (i.e. they picked a small bit of the growth from the first culture and put it onto a fresh tube of the same medium), and they repeated this every three weeks for 13 years. This would have been a daunting prospect for any microbiologist; the fact that they continued right through the First World War makes it an astonishing achievement. But it worked, in that BCG is a very safe vaccine (although, as mentioned in Chapter 3, it is not clear how well it works).

There was a lot of controversy for many years over whether the vaccine really was safe to use, and whether it would revert to being virulent again. This was not helped by an unfortunate incident in Lubeck in Germany, in 1930, where 72 children died of TB after receiving the vaccine. It was subsequently established that this was due not to the vaccine reverting to virulence, but to a disastrous error in making the vaccine, where a virulent TB strain, rather than the attenuated one, was used by mistake. In a subsequent section, I will look at how vaccines are tested for safety and efficacy.

One problem with live vaccines is reversion, as we saw in Chapter 4 with the polio vaccine. The live polio vaccine was developed by repeated culture of the virus, at low temperatures, in monkey kidney cells. During this process, it became attenuated due to simple point mutations, i.e. a change of one base in the genome. However, a reversal of that mutation can restore virulence. In principle, this could be prevented by introducing several different mutations, or by deleting longer stretches of nucleic acid, but this can be difficult with viruses; they have a relatively small number of genes, so too many changes can make the virus impossible to grow in the lab. It is easier to introduce such changes into a bacterial strain. For example, reversion does not happen with BCG, because there are too many differences between BCG and the virulent bacterium, including the complete deletion of some regions of the DNA that are thought to be necessary for virulence. Deletions will not revert; the bacterium cannot re-acquire large stretches of DNA.

Nowadays, we require vaccines with a greater degree of attenuation than was the case in the past (notably with the smallpox vaccine), but this can be difficult to achieve. Using conventional methods, it requires an exhaustive process of mutation and selection to obtain a strain that is sufficiently attenuated but is still capable of establishing a protective immune response. One of the problems here is that the pathogen may become *too* attenuated. The best response to a live vaccine occurs if it is able to survive, and perhaps replicate in the body to a limited extent, so that there is a chance that the body will recognize it and make antibodies (or immune cells). It can be difficult to get the right balance. Too much attenuation means that it is eliminated too rapidly, without this happening. Too little attenuation means that it is not a safe vaccine. Thus, vaccinia, for example, is a very effective vaccine against smallpox, but it is not a very safe vaccine as it can cause problems, especially in people who are immunocompromised, or who suffer from certain skin conditions. On the other hand, BCG is a very safe vaccine, but it could be argued that its limited effectiveness indicates that it is over-attenuated.

It can, therefore, take a lot of research effort, and time, to produce a safe and effective live attenuated vaccine. Making a killed vaccine is in general much simpler. In principle, all that is necessary is simply to grow the pathogen on a large scale and then inactivate it, for example by treatment with heat or formalin. Of course, the method of inactivation has to be chosen carefully so that it does not destroy the immunogenicity of the vaccine, and it has to be completely effective, so that no live pathogen remains.

A killed vaccine can be especially useful if a vaccine is needed quickly – for example, when a new strain of influenza virus starts a pandemic. However, for influenza, even a killed vaccine cannot be produced very quickly, because the virus is not easy to grow in the laboratory. The usual procedure is to use embryonated hens' eggs (fertilized eggs containing a growing chick embryo). A small part of the shell has to be carefully removed, leaving the underlying membrane intact, and the virus is injected through that membrane. After incubation of the infected eggs to allow the virus to grow, the contents are harvested and the virus extracted. It then has to be purified to remove the egg proteins, and then tested to make sure it is safe to use, as well as determining and standardizing the amount of virus it contains. It takes three eggs to produce one dose of vaccine, so the 300 million doses produced per year require nearly a billion eggs – and a billion embryonated hen's eggs cannot be produced at the drop of a hat! Thus, it takes time to make this vaccine on a large scale.

In some cases, even a killed vaccine is not safe to use. This is especially true for bacterial vaccines, where the whole cell extract may contain several products that are themselves toxic. One way around this is to treat the cell extract to remove any toxic products, leaving just the components that are needed for the vaccine. This is called a subunit vaccine. One example, the vaccine against pertussis (whooping cough), was described in Chapter 4.

A second example is the vaccine against *Haemophilus influenzae* type B (the Hib vaccine). Despite its name, this bacterium does not cause influenza, but is a cause of meningitis and other serious infections in young children. For its virulence it requires an external coat (a capsule), and immunity to that capsule confers protection against the disease. However, this capsule is made up of polysaccharides, which are poorly immunogenic, especially in very young children whose immune system is not fully developed. To get around this problem, the polysaccharide is chemically linked to (conjugated with) a protein carrier such as the tetanus toxoid (see below). The use of this vaccine has been highly successful in preventing *Haemophilus* meningitis in young children.

Toxoids were also briefly introduced in Chapter 4. These are relevant for diseases such as diphtheria and tetanus, where the symptoms of the infection are due to a single toxic protein. By purifying the toxin and then inactivating it, it is possible to produce a simple vaccine that will confer immunity to that disease.

Diphtheria, which is caused by the bacterium *Corynebacterium diphtheriae*, used to be a significant cause of deaths in children (and still is in some parts of the world). The symptoms are due to a protein toxin that disrupts protein synthesis in the affected cells, which causes the production of an exudate that clots, producing a sort of membrane across the throat that can block breathing. To produce the vaccine, the toxin is purified and treated with formaldehyde, which creates links between some of the amino acids in the protein chain. This makes it non-toxic, but it remains antigenic. Receiving the vaccine causes antibodies to the toxin to be made so that, if there is a subsequent infection by this bacterium, the body neutralizes the toxin and, hence, no symptoms result.

Tetanus was described in Chapter 5. The causative bacterium, *Clostridium tetani*, is an anaerobic bacterium, so it will only grow in the absence of air, but the conditions in a wound, especially if some dirt or other foreign bodies are present, allow the bacterium to multiply and produce a protein toxin which affects the nervous system, causing the muscles to contract uncontrollably. As with the diphtheria vaccine, purification and inactivation of the toxin produces a toxoid, which is a highly effective vaccine for prevention of tetanus.

The production of killed or subunit vaccines in the conventional way requires the growth of large amounts of a virulent pathogen, which is inherently dangerous and, therefore, needs extensive (and expensive) safety measures. In addition, manufacturers obviously have to make sure that the virus or bacterium has indeed been completely killed, so that the vaccine is safe to use. Recombinant DNA technology (or 'genetic engineering') opens up new approaches to the production of vaccines that avoids these problems, as well as providing totally new ways of making vaccines.

The first of these approaches is to clone the appropriate gene (as described in Chapter 7) in a non-pathogenic bacterium (or other convenient microbe), so that we can use these safe microbes to produce a protein for use as a vaccine, rather than growing up large amounts of the pathogen. The best example of this is hepatitis B. This virus has proved impossible to grow in the laboratory, so early attempts to make a vaccine relied on purifying the virus from the blood of people with the disease. Not only were there severe limitations on the supply of virus by this route, but also major safety implications. For example, we have to take into account the fact that the main routes for infection with hepatitis B are unsafe sex and the use of contaminated syringe needles – which are also the main routes for transmission of AIDS. Thus, there is a risk that the source of the material (a person infected with the hepatitis B virus) may also be infected with HIV.

We know that one of the proteins of the hepatitis B virus coat, the surface antigen, is able to induce protective immunity. It was therefore thought that cloning this gene, inserting it into *E. coli* and getting the bacterium to express the gene would provide a convenient source of the protein for vaccine production. Unfortunately, it was not that simple. The gene was expressed, but the protein that was made did not adopt the right shape and, hence, was unable to induce a proper immune response. This problem was overcome by using an alternative host, the yeast *Saccharomyces cerevisiae*, instead of *E. coli*. *S. cerevisiae* is the common yeast used for brewing and baking, so it is easy to grow in large amounts, and this proved to be an excellent source. The vaccine now in use for hepatitis B is derived in this way.

A similar approach can also be used to produce novel live vaccines, by taking genes from a pathogen (or several different pathogens) and inserting them into an existing viral or bacterial vaccine strain. In this way, a live vaccine can be made that will confer immunity to several different infections. The smallpox vaccine, vaccinia, has been used as a carrier for this purpose, as have some bacterial vaccines, such as BCG.

A very different use for gene technology is in the production of specifically attenuated pathogens. Rather than inserting new genes, it is possible to remove existing ones (known as a 'gene knock-out'). To explain this, we have to remember, as described in Chapter 7, that there is a natural tendency for similar bits of DNA to pair together and recombine, by breaking and re-joining the DNA, so that parts of the DNA are swapped. If we make, in the lab, a gene that doesn't work (e.g. it might have a bit missing, or a few bases changed) and put that into a cell, then it can, by recombination, replace the proper gene. We thus get a gene knock-out.

Even more radically, we can dispense with the organisms altogether and just use specific bits of DNA from the pathogen in question. Rather surprisingly, it was found that the injection of DNA rather than proteins could, using the right combination of materials, produce protective immunity at least in animal models. Apart from the hepatitis B vaccine, most of these approaches have yet to come into widespread use for human infections, but we are likely to see products emerging in the near future.

However we make a new vaccine, there are obvious questions to be addressed before it is introduced – is it safe, and does it work? If there is a good animal model available – that is, an experimental animal which can be infected with the pathogen to produce a disease similar to that in humans – then it can be seen whether the vaccine has lost the ability to cause disease in those animals. The animal model can also be used to test whether the vaccine works. To do this, we could start with two groups of animals. One group is vaccinated and the other group (the control group) is not vaccinated. These animals can then be challenged by giving them a suitable dose of the pathogen. The limitation is that, for many diseases, the animal models are not very good mimics of the infection in humans.

Another approach (in humans or animals) is to use what are called 'correlates of protection'. For example, if it is known that production of antibodies to a key antigen of the pathogen is a good indicator of immunological protection, then the levels of that antibody in the vaccinated and control groups can simply be measured. If the vaccine results in good antibody levels, then it is probably a good vaccine.

However, for many diseases, the reasons why a vaccine is protective are not well understood, especially if protection relies on cellular immunity rather than antibodies. Vaccines often produce a variety of different immunological responses, some of which are protective and some are not. Unless it is known what the correlates of protection are, this approach is not possible.

Ultimately, however good the preliminary tests are, the vaccine will need to be tested on humans.

The first stage, known as Phase I trials, is carried out to see if there any side effects in a small number of healthy volunteers. If the vaccine passes this stage, it moves on to a Phase II trial, with rather larger numbers.

At the Phase II stage, we are interested in further assessment of safety as well as the nature of the immune response. This trial may also evaluate factors such as the dose that should be given and the route of administration. If the results are good enough, it may be worth moving on to a full-scale Phase III trial.

Phase III is the really expensive bit. As with the animal tests above, the test population is divided into two groups. One group receives the vaccine, the other does not. The difference is that we obviously cannot challenge the subjects by deliberately infecting them. We have to wait and see whether there is a significant difference in the level of disease in the two groups. This is simple in principle, but more complicated, and more expensive, in practice. Unless the disease is very common, there might need to be tens of thousands, or even hundreds of thousands, in each group, and it might be necessary to follow those groups for several years to see whether there is an effect.

The Phase III trial will also provide a more realistic assessment of the safety of the vaccine, but this is less straightforward for two reasons. Firstly, what side-effects are we going to look for? Some are possibly obvious, but what about unexpected side-effects? Detailed questionnaires or interviews could be used to ask about all aspects of a subject's health before and after immunization, but this is unlikely to be feasible in a large-scale trial involving tens of thousands of people. However, a detailed study of a smaller number of people is likely to miss any rare side-effects – and this brings us to the second limitation. Even a large-scale trial will involve far fewer people than are involved in a mass vaccination campaign. The only way around this is to define a minimum level of risk that can be regarded as acceptable, and make the trials big enough to be able to say that the level of identified side-effects is below this level. Of course, it is possible – indeed, essential – to re-evaluate the risks once the vaccine is in use, by monitoring the extent and nature of possible side-effects.

This raises uncomfortable issues, as there is constant media and public demand for statements that a vaccine is 'safe', and this demand is not satisfied by responses that the level of risk is less than one in a million (or whatever). A similar situation arises with new drugs and other medical interventions, as well as with food safety. Yet, in other situations, we are quite ready to accept that nothing is 'risk-free' – crossing the road, driving a car, even getting out of bed in the morning. We are prepared to accept a certain level of risk in all our activities, because otherwise life would become impossible. It is a similar case with vaccinations; we can demand that the risk is reduced to a certain level, and beyond that we should accept that the possible risk is justified by the enormous benefits created by the vaccine. The question of the assessment of risk, and the controversies regarding the safety of vaccines such as MMR, are considered further in Chapter 10.

9.8 Proteins

In Chapter 7, I said that it is possible, using gene cloning, to get microbial cells to produce foreign proteins, thus providing a cheap and easy way of obtaining large quantities of the protein. We have already encountered one example – the hepatitis B surface antigen – in this chapter. In principle, this can be done with any protein, from any source – other microbes, plants, animals, or wherever. Once the gene responsible has been cloned, appropriate bits of DNA can be attached to provide expression signals, such as a good promoter that will drive a high level of transcription (see Chapter 7). With luck, the recombinant bacterium (usually *E. coli*) can be persuaded to make a lot of protein – maybe up to 50 per cent of the total protein of the cell. Since this organism can be easily grown on a large scale, virtually unlimited amounts can be obtained. Not only does this provide a prolific source of that protein, but the fact that there is such a lot of it makes downstream processing (i.e. separating the specific protein from other material that we don't want) much easier and more efficient.

However, there are problems arising from producing such a large amount of protein that is of no use to the cell. First of all, it will grow more slowly, which is clearly undesirable in itself. Worse than that, any mutant that no longer makes the protein (perhaps because the gene has been deleted, or the plasmid that carries the gene has been lost) will grow much faster and will take over the culture. On a laboratory scale, this is merely a nuisance, and it can be partially overcome by including antibiotics in the growth medium, so that any cells that lose the plasmid will die. Doing this on an industrial scale, though, would be both impractical and uneconomic; millions of litres of spent medium containing the antibiotic would be produced, and getting rid of that would be a real problem.

There are various tricks that can be used to cope with this situation. The main one is to use a promoter that can be turned on or off by changing the conditions. For example, if an inducible promoter is used (one that will not be active unless a specific chemical is added), the culture can be grown on a large scale in the absence of that inducer. Under these conditions, the cloned gene will not be expressed, so the bacteria will grow well. Then, when there is enough growth, the inducer can be added to turn the promoter on, thus producing a massive burst of expression of the cloned gene (this is not actually the best way of doing it, but it is the simplest to describe and it illustrates the principle).

There are an enormous number of applications of this concept. Many of the enzymes that are used in molecular biology, some of which we have already encountered (e.g. restriction enzymes and DNA ligase), are produced in this way. It also provides a good source for human enzymes that are used therapeutically, the earliest example being human plasminogen activator, popularly known as a 'clot-buster', as it is used for treating heart attacks caused by a clot blocking a coronary artery. Other enzymes made by recombinant microbes include the rennet used for cheese making, and the enzymes that are added to biological washing powders.

This is not limited to enzymes. Many other proteins, including peptide hormones, can also be made by genetically engineered microbes (a technical note: I refer to these as peptides rather than proteins, as they are smaller than proteins, but they are still coded for by DNA, which can be cloned and expressed in the same way). The first major success was human growth hormone, used in the treatment of pituitary dwarfism. Children with this condition, which prevents normal growth, were previously treated with extracts of human pituitary gland, extracted post-mortem.

Not only was the supply extremely limited and difficult to obtain, but there were serious health implications, not only for the patient but also for the staff involved in extracting and purifying it. Indeed, the risks were greater than was appreciated at the time, as it was not then known that some diseases (especially the spongiform encephalopathies such as Creutzfeld-Jacob disease) could be transmitted by such material. The cloning of the human growth hormone gene, and its expression in *E. coli*, therefore represented a major step forward. For the first time, it provided a cheap and safe source of this drug, in unlimited quantities.

A second famous success was the cloning and expression of the human insulin gene. The usual source of insulin for treating diabetics was originally pigs, but pig insulin is slightly different from human insulin. This caused problems in some people, and there is always a risk of transmission of unknown pathogens when using material from non-human sources.

In principle, it is not necessary to restrict microbial production to naturally occurring proteins. Once a gene has been cloned and its sequence determined, it is quite easy to introduce changes into the DNA sequence so that the structure of the protein is altered in a specific way. For example, one limitation to the commercial use of enzymes is that they tend to be heat-sensitive, so they can only be used at moderate temperatures. If the enzyme can be made more heat stable, then the process can be run at a higher temperature, which is more effective. Thus, it can be useful to change the structure of the enzyme to extend the temperature range over which it remains active. To appreciate how this is done, it is necessary to understand a bit more about how enzymes work.

Proteins are chains of amino acids, but they do not have a linear structure. Interactions between the amino acids make it fold up into a specific shape. Within that shape, an enzyme has a 'pocket', which is a hole in the structure which fits the substrate (the chemical that the enzyme is to act on). The amino acids lining the pocket then interact catalytically with the substrate, which makes the reaction work (see Appendix 1 for more information). Thus, the activity of an enzyme depends on the way in which it folds up on itself.

However, the interactions that give the protein its shape are relatively weak, and they will tend to fail if the protein is heated. Therefore, if the temperature is raised, the enzyme will tend to unfold and so lose its activity. If chemical links between the chains can be introduced, this can make the structure more stable, and it can be done by changing the sequence of amino acids – specifically by incorporating cysteines into the sequence. Cysteine is a sulphur-containing amino acid which tends to form stable crosslinks between two parts of the amino acid chain. These crosslinks stabilize the structure and make it more heat-stable.

Other changes can be introduced that make the enzyme less sensitive to chemicals that would otherwise interfere with its action. More subtle changes can be made that will change the specificity of the enzyme, so that it will attack new substrates, or lose the ability to attack substrates that we do not want it to attack. Even more radically, it is at least theoretically possible to design completely new enzymes. Rather than changing an existing gene, we can synthesize a totally new one from scratch and thus make a new gene that codes for a completely novel protein.

This field, known as *protein engineering*, potentially offers a wealth of possibilities, but it has to be said that in practice it is not as easy as it sounds. The main limitation is that we do not fully understand the relationship between the shape of the protein and its activity. For example, the activity of an enzyme depends not only on its shape, but also often on the ability to alter that shape in response to the presence of a substrate. It may need to be flexible so that, in effect, it curls around the substrate and amino acids from different parts of the protein make contact with the substrate in order to carry out the chemical reactions involved. Thus, changing one or more amino acids may have unexpected consequences for the activity of the enzyme. This is most clearly exemplified by the introduction of cysteine residues as described above. They are certainly likely to make the protein more heat-stable, but they may also make it too rigid, and hence it may lose its enzyme activity.

I have now covered a range of topics that illustrate the significance of microbes in the natural world and how we can use microbes for our own ends. In the final chapter, I want to look at some topics that are controversial, and introduce some speculations about areas that remain unanswered.

10
Controversies and Speculations

The purpose of this chapter is to explore some topics that are controversial (justifiably or not) and to introduce a bit of speculation on others. It also gives an opportunity to develop some matters a little further.

10.1 Evolution and the origins of life

The 'Theory of Evolution' (as introduced in Chapter 7) is a contentious topic, at least in non-scientific circles. In part, this is because of the ambiguity of the word 'theory'. In everyday use, 'theory' is usually taken to mean a speculative idea that needs to be tested, or proved. In a crime novel, the detective may have a theory as to who committed the murder, and sets about gathering evidence to support that theory. In scientific language, however, we would describe that as a hypothesis, not a theory, and we would test it by making predictions based on the hypothesis. If those predictions are wrong, then the hypothesis needs to be discarded or, at least, modified. Scientifically, a 'theory' is quite different. It means an overall framework of ideas that accounts for the known facts. Such is the Theory of Evolution.

The most remarkable thing about evolution is why it should be so contentious, when it is obvious from first principles, as can be seen by breaking it down into simple, and quite intuitive, concepts.

Firstly, variation exists. Different individuals of the same species are not identical. This is most obvious with domesticated animals – dogs or cows, for example – but it can also be observed in wild animals and plants, and can be seen if we look closely enough (the nature of those variations, and how they happen, was dealt with in Chapter 7).

Secondly, some of these variations will be helpful, while some will be unhelpful. Some colour variations are useful in camouflaging an individual, thus making it

---

Understanding Microbes: An Introduction to a Small World, First Edition. Jeremy W. Dale.
© 2013 John Wiley & Sons, Ltd. Published 2013 by John Wiley & Sons, Ltd.

less susceptible to attack by predators. Therefore, there is a selective pressure in favour of certain variants, which will grow better, survive better and, above all, reproduce better. Hence, the proportion of the population with that variation will increase generation by generation. Other obvious examples would be the ability to survive drought, or to tolerate hot or cold conditions. Different variants will come to predominate in different environments.

This leads on to the third principle, that of isolation. If the organisms are moving around and mixing freely, they will remain as variants within a mixed population. Some will do better in one place than another but, because they are interbreeding, we will just see a spectrum of variation. However, if they are geographically separated – for example, on two islands that are far enough apart to restrict movement between them – they will tend to diverge and become more and more specialized for the conditions that prevail where they live. What is more, the isolation does not have to be geographical; any factor that reduces the extent of interbreeding between two variants will have the same effect. For example, two variants of a plant species that flower at different times will not fertilize one another (in the wild anyway). Since there is no interbreeding, further variation will occur within each, until eventually we would recognize them as different species.

So, we have developed a theory of the origin of species that is based on a few intuitively obvious premises: natural variation; selection for variants that are better adapted to their environment; and isolation which allows the variation to accumulate to an extent that they can be labelled as different species. If it is so obvious, why did the publication of Darwin's *On the Origin of Species by Means of Natural Selection, or the Preservation of Favoured Races in the Struggle for Life* (its full original title) cause such a fuss when it was published in 1859? It is worth noting that it did not come totally out of the blue. Ideas about the gradual evolution of species, and opposing views, had been circulating for many years before that. Most of these were either vague or dogmatic, and were not based on good science. Above all, there was a tendency to avoid the thorny problem that these ideas would apply to human evolution as much as to the evolution of other species.

Darwin's achievement, not forgetting that of Alfred Russel Wallace who independently put forward the same concept, was in putting together a coherent set of ideas based on his study, over a number of years, of an enormous number of examples of the process – famously that of the finches in the Galapagos Islands with their characteristic differences in the shape and size of their beaks.

You may object that it is all very well talking about relatively simple features such as colour or beak size, or even the long necks of giraffes, which provide a clear advantage in being able to reach leaves that are out of reach of other animals. What about more complex features? Eyes are often quoted as an example; how can such a complex structure evolve by gradual accumulation of simple variation? Well, quite easily, actually, if you remember to factor in the enormous length of time and, hence, the large number of generations over which evolution has occurred.

Figure 10.1   Darwin's finches. Darwin's drawings of the beaks of finches from the Galapagos Islands. (*Source:* see http://darwin-online.org.uk).

Suppose we start with a basic single-celled organism that cannot detect light. The membrane surrounding the cell contains many proteins that recognize conditions outside the cell, such as the presence of a specific food. These proteins respond by changing their shape, which transmits a signal to the inside of the cell to alter the expression of relevant genes. Imagine a simple change in a gene coding for one of these membrane proteins so that the protein responds to light rather than to a specific chemical. The protein now alters its shape when exposed to light, with the result that the cell will only switch on its photosynthetic machinery if there is enough light around. There is an evolutionary advantage here, so such a variant would tend to proliferate. If a further change occurs so that it can sense the direction the light is coming from (e.g. by having this protein at the bottom of a pit on its surface), then it can swim towards the light. This has now become a rudimentary eye.

Further variations may then allow this protein to occur in a series of discrete spots within the pit, so producing a crude image, and this is followed by further changes that result in a membrane over the surface, allowing focusing of the light to get a better image. At each stage, there is a further advantage for that variant. Eventually, after millions of years, and through the accumulated effect of a very large number of small changes, we get the enormous variety of different structures that we call 'eyes'.

Of course, Darwin did not know about the molecular nature of evolution, nor how variation occurred. Even the nature of a gene was still to be developed (the word seems to have been first used in 1909 by the Danish botanist Wilhelm Johannsen). Consequently, some of the concepts in *The Origin of Species* have been modified and adapted since, but the basic underlying idea remains and has stood the test of time. It supplies a completely adequate explanation for the origin of the enormous variety of species that currently exist – microbes, plants, animals – starting from an initial simple cell. There are, of course, a number of details that still need to be pencilled in, such as how quickly it happened, whether it proceeded smoothly or in fits and starts, and exactly where the branch points are – in other words, what was the most recent common ancestor for each of the species we see today? These details are being gradually filled in, but they do not affect the overall message.

The one big question that remains unanswered is the nature and origin of that first simple cell. How did it all start? Even the simplest sort of cell that exists today, for example a basic bacterial cell, is quite complex. It has DNA and machinery for replicating it, enzymes that copy the DNA into RNA, the ability to make proteins using an RNA template (including ribosomes and transfer RNAs), and a cell membrane which not only separates the cytoplasm from the environment but also controls the uptake of food and secretion of products into the environment. These factors are common to all bacterial cells. It seems inconceivable that such a complex structure could have originated in its complete form by chance, even within the timescale of geological time and the extent of the environment available. So where did this first cell come from?

I will straight away discount the hypothesis that it arrived on Earth from somewhere else in the universe – not because it is impossible (although I think it is unlikely) but because it does not answer the question. If it arrived from somewhere else, how did it develop there?

The fundamental problem lies in the assumption that the initial precursor cell must have looked rather like something that exists today. This is not necessarily true. Let us look at a very tentative and speculative model of how things might have happened.

Consider the situation on Earth before life emerged. The so-called 'primordial soup' might well have contained, in tiny amounts, many of the basic chemicals needed for life to start. Modern experiments have demonstrated that many of these chemicals, such as amino acids, can be formed abiotically (in the absence of life), aided by high temperatures and pressures, and a highly reactive environment compared to the situation we observe in most parts of the world today. Within this soup, it is not inconceivable that tiny amounts of something resembling modern RNA could have formed. I'm only thinking of very short sequences, but they could have the important property of promoting (very inefficiently and inaccurately) the assembly of copies of themselves by forming base pairs, which would aid the (non-enzymic) joining together of the nucleotides. These are random bits of RNA, but eventually the process of inaccurate replication would evolve RNA molecules that had some catalytic ability, so they would copy themselves more effectively. Again, we can get some support from modern experiments which have shown that RNA does have some catalytic ability, and also that replicating RNA molecules can evolve abiotically.

Since we are imagining all this happening within the 'soup', without any sort of compartmentalization, it would be extremely inefficient. So let us now imagine some sort of membrane forming around our replicating RNA. It is not difficult to imagine this happening by chance. If there happened to be some fats (lipids) in the 'soup', they would tend to form globules, because they are hydrophobic. Given the right sort of lipids, these globules can contain some of the chemicals in the vicinity – a property that is commonly exploited today in the formation of what are called liposomes (which are used, for example, to help get various drugs into their target cells – see Appendix 1).

Now we have imagined something that could be regarded as the precursor of a cell, although still lacking many of the features we regard as necessary – no DNA, just RNA, and no proteins or protein-synthesizing machinery such as ribosomes and transfer RNA. The absence of proteins means the absence of enzymes (apart from the catalytic activity of the RNA). Enzymes are so central to all the activities of a 'modern' cell that we tend to regard them as essential for those activities – but they are not magic. They cannot make a chemical reaction happen unless that reaction is already a possibility. They make a reaction go faster, and their specificity ensures that metabolism is not just a set of random reactions, but it is possible to imagine a highly inefficient primitive 'cell' operating without enzymes.

I do not propose to go through all the steps by which these additional features might have developed, except to say that it is quite easy to imagine an incremental, step-by-step process, in which chance events just occasionally happen to produce an improvement on what went before, so that 'cell' would be more efficient than its predecessors and thus would be selected. A lot of blind alleys would be explored and discarded along the way. So, this imaginary process exploits the same concepts as those developed by Darwin – random events, some of which lead to an advantageous property which therefore result in selection of a modified life form.

Of course, all this is just an armchair model, and there are many other models that could be (and have been) developed. All of these imaginary precursors would have been so ineffective by modern standards that they would have disappeared millions of years ago, without leaving even a fossil trace of their existence.

However, there is one aspect of the development of cells that did *not* proceed by such a gradual process, and that relates to the development of eukaryotic cells (those with a nucleus containing the genetic material). Life as we know it appears to have originated on Earth about 3.8 billion years ago, and for nearly two billion years there were only prokaryotic cells. Then, eukaryotes appeared. Apart from the nucleus, these differ from prokaryotic cells in having organelles such as mitochondria, which are often described as the powerhouse of the cell, as they are largely responsible for energy-generating processes. Photosynthetic eukaryotes also have chloroplasts, which are responsible for photosynthesis. These organelles, mitochondria and chloroplasts, are intriguing, as they have their own DNA and are surrounded by a membrane. Where did these organelles come from?

In 1970, Lynn Margulis made the radical suggestion that mitochondria and chloroplasts did not develop from scratch, by the normal gradual process of mutation and selection, but happened in a more dramatic way, when one cell engulfed a smaller one. The engulfed cell then became adapted to life inside the larger cell, losing much (but not all) of its original DNA but retaining the ability to make proteins and generate energy (or, for chloroplasts, photosynthetic ability). Experimental evidence was provided in 1978 by Robert Schwartz and Margaret Dayhoff, who showed (using DNA sequence data) that chloroplasts share a 'recent' ancestry with photosynthetic cyanobacteria, and mitochondria with a group of bacteria known as the Rhodospirillaceae. This means that these organelles are actually symbiotic microbes, or at least the remains of them. The acquisition of mitochondria and chloroplasts must have occurred after the development of a nucleus in the host cell, and it is tempting to suggest that the nucleus appeared in the same way, but here the evidence is not strong.

10.2   Is there life elsewhere in the universe?

In 1996, NASA scientists announced that they had found evidence of life in a meteorite that had been found in Antarctica. This meteorite (ALH 84001), which was believed to have originated from Mars and travelled through space for millions of years, crashed into Earth 13,000 years ago. The claim caused extensive controversy, and it is now widely believed that the structures seen were not of biological origin. Nevertheless, the search continues for evidence of life elsewhere in the Universe.

If we believe the speculation in the previous section (or something resembling it) – that life on Earth originated through a succession of random chemical events – then there must be a good chance that something similar would have happened, at some time, somewhere else. This is especially so when we factor in the vast number of stars and the many planets that are now being identified orbiting them. But what would that life be like? Comics and films about creatures from outer space tend to make the aliens look remarkably like ourselves – maybe with a different number of arms and legs, or even heads, and perhaps green or purple – but still recognizable. Even if we discount science fiction, serious attempts to find life elsewhere tend to assume that the underlying biochemistry will be similar – so the focus, at least in the media, is on finding evidence of water or oxygen to sustain 'life', as well as organic chemicals such as amino acids that would indicate biological activity. Based on this assumption, the search is for planets that are a similar size and temperature to Earth, mainly because it is only under such conditions that liquid water will exist.

Perhaps we need to be more imaginative. We tend to assume that life arose on Earth because the conditions were suitable for it, whereas maybe life on Earth is the way it is because it has evolved to suit the conditions. And even so, the conditions are extremely variable. As we saw in Chapter 6, some organisms thrive under extremes of temperature, pressure and pH, and with or without oxygen. If life on Earth, all of which appears to have originated from a single ancestor, can be so varied in its requirements, is it not possible that a form of life on another planet, originating independently, could be radically different? Can we imagine a form of life that does not require water?

· It is reasonable to assume that there must be some sort of liquid available – it is difficult to conceive of a life form existing solely in a gas or solid phase. But does the liquid have to be water? At high pressures and low temperatures, various substances that exist as gases on Earth, such as methane or ammonia, are liquids. Titan, Saturn's largest moon, has lakes of liquid methane. Maybe somewhere there, or elsewhere, there are organisms thriving in pools of methane rather than water.

10.3   Creating new life

Implicit in much of the previous sections is the challenging concept that there is no clear boundary between 'living' and 'non-living'. If a living cell can originate through a (very extensive) series of chance events, from non-living starting material, then is 'life' any more than a complex set of biochemical reactions? It is decidedly uncomfortable to think that this is all we are, and it impinges greatly on philosophical and religious questions that I would rather not go into.

But if this is the case, then is it possible to create a living being in the laboratory, starting from a set of chemicals that we can buy? We know that it is certainly possible, and has been done, for some viruses. It is relatively straightforward to synthesize a complete viral nucleic acid (DNA or RNA, depending on the virus) and, if that is put into an appropriate cell, it results in infectious virus particles. However, making viral DNA is not the same as 'creating life'. Although virology is conventionally included as part of microbiology, there is a good case for saying that viruses are not really 'living'. After all, we have to put the nucleic acid into a living cell to get viral particles made. Can we do the same thing with a real living cell?

In 2010, a major step in this direction was carried out by a team led by Craig Venter in the USA. They synthesized from scratch the entire genome of a simple bacterium, *Mycoplasma mycoides,* and inserted this into a cell of the related species *M. capricolum*, from which the native DNA had been removed. This cell was then able to grow, and it took on the properties of the organism for which the DNA had been synthesized.

At the time, this was proclaimed as 'creating new life' whereas in reality it was no such thing. In essence, it was not very different from the synthetic viruses referred to above, in that it still used a 'living' cell to copy and use the DNA that had been made. However, it does open up the way to creating novel species of bacteria, as it is quite possible to make changes to the DNA – leaving out some genes and inserting others – so as to make an organism with totally new properties. This is still quite a long way from creating a new cell from scratch, though.

10.4   Is it safe? Assessment of risk, risk versus benefit

I now want to move onto a different area of controversy, which relates to the safety of vaccines, following on from Chapter 9, where I looked at the general question of vaccine safety and efficacy, and how vaccines are tested.

One example, that of the pertussis (whooping cough) vaccine, was described in Chapter 4. In the 1970s, there were claims that some children developed fits and neurological damage after receiving the vaccine. As a consequence, some parents declined to have their children vaccinated. The predictable result of the lower vaccine uptake was epidemics of pertussis, with many deaths and cases of severe neurological damage arising from the disease. This was doubly unfortunate, as the doubts about the safety of the vaccine were misconceived. Undoubtedly, fits did occur in some children after receiving the vaccine, but it cannot be said that these were *caused by* the vaccine. Some children of that age do have fits anyway, so the fact that some had fits soon after having the vaccine does not mean that the vaccine was the cause.

In order to test this, it is necessary to compare two groups of children – one group receiving the vaccine, and a control group who are not vaccinated – and then see how many in each group suffer fits. Studies carried out like this have not shown any difference in the frequency of fits (or other complications) between vaccinated and non-vaccinated children.

A similar, more recent, story relates to the MMR (measles mumps rubella) vaccine. The controversy here started in 1998, when Andrew Wakefield and colleagues studied 12 children with autism and chronic intestinal symptoms, with the implied hypothesis that there was an association between intestinal disorders and autism. As an incidental outcome of that study, they found that most of these children had received the MMR vaccine. Although the paper actually stated explicitly that '*We did not prove an association between [MMR] and the syndrome described*', it was widely reported as indicating such an association, partly on the basis of press statements by Wakefield.

During the succeeding years, a number of large and well-designed studies were carried out in an attempt to ascertain if there was such a link, and none was found. These studies received little attention in the media – 'Epidemiological study fails to find a link between MMR and autism' does not make a good headline, and the only honest answer to the question 'Does MMR cause autism?' is that there is no evidence that it does. As a consequence of the media continuing to fuel this unwarranted speculation, the level of uptake of the MMR vaccine fell well below the required level (down to 80 per cent in England by 2003) – with the further consequence of epidemics of measles and mumps, reaching over 40,000 cases of mumps in 2005.

Why has it been so difficult to convince people that MMR is safe? One factor is that autism is, to a large extent, an unexplained condition, and people understandably want explanations. Therefore, they look for things that happened to their child before symptoms of autism were noticed. In eight of the 12 cases studied by Wakefield, the parents had associated behavioural changes with MMR vaccination. Anecdotal evidence of this sort can sometimes provide useful indicators of where to look for a cause, but it does not provide proper evidence by itself. The fact that one event followed another does not mean that the first was the cause. This morning, I had toast for breakfast and then started work. But it wasn't the toast that *caused* me to start work; these are two events that generally occur at that time of day. Similarly, the age at which MMR is usually given is similar to the age at which autism is first diagnosed, so it is no surprise that autism is sometimes detected in children who have recently received MMR.

Three features of the Wakefield paper are worth drawing attention to:

- firstly, the small scale of the study (only 12 children);

- secondly, this was a self-referred group, not a random sample, so they were investigated because of the presence of autistic symptoms;

- and thirdly, the variable length of time between vaccination and the identification of symptoms. In the Wakefield paper, this interval was reported as up to 14 days, but that only refers to the eight children whose parents had associated the behavioural problems with MMR. It does not include three children, whose parents did not report an association with MMR, where vaccination occurred 1–2 months before behavioural changes were noticed.

So, the statement in the paper is not actually untrue, but it is a highly selective view of the data.

A properly designed study can test whether there is a statistically significant association between such events. Technically, this is done by setting up what is called a 'null hypothesis' (i.e. a hypothesis that there is no such association) and analyzing the data to see how likely the null hypothesis is). To anyone not familiar with statistics, this may seem an odd thing to do, but it is actually quite logical on the basis that statistical methodology – and, indeed, scientific methodology in general – is directed at testing and rejecting hypotheses. In this case, if the two events are unconnected, there is still a chance that sometimes one will follow the other, because that is how chance works.

The statistical analysis will tell us how likely it is that the results could have happened by chance. If this is less than one in 20 (formally, we would say that the probability, p, is less than 0.05), then we would reject the null hypothesis and say that there is an association; it is unlikely to have happened by chance. There is still, of course a one in 20 chance that the association does not really exist so, if we were to carry out 20 such studies, we would expect (on average) one of them to show results like this. If we get an even lower p value – say, less than 0.01 – then the evidence for an association is even stronger. At $p = 0.01$, the chance of getting such a result without a real association is one in a hundred. We can draw the line at different values, but 0.05 is the highest value that would be accepted as significant.

If p is greater than 0.05, we cannot reject the null hypothesis – in other words, the data that we have could have originated just by chance. The difficulty is that this does not prove that there is no association – simply that we do not have any evidence for one. However, if the studies are large enough, and if there are several such studies (as is the case here), then we can be confident that the effect does not exist.

But we can never say that the data proves that there is no risk – only that any risk must be smaller than could be detected by the study we have carried out. Hence the difficulty that scientists have when faced with the question 'Is it safe?' The honest answer – that we have been unable to detect any risk – tends to look as though we are equivocating, and the media, the politicians and the public find this unsatisfactory.

At the risk of getting too technical, it is worth looking further at the sort of data we would get from a study of vaccine safety, and how the data would be analyzed. This does not just apply to vaccine safety – the same procedures would be used for assessing other risks, such as those that might be associated with eating certain foods, living near nuclear power stations or taking specific drugs. Or we might be trying to test whether a new disease is caused by a specific microbe, or what other factors are involved, ranging from smoking to genetic differences.

If we stick to vaccine safety as our example, we could do the study as described in the previous section – divide people into two groups, one group being vaccinated and the other (the control group) not vaccinated. Then, after an appropriate time interval, we determine how many people in each group have developed the symptoms we are looking for. Before considering how that data should be analyzed, it has to be said that a study like this (a cohort study) is not always possible. If the outcome we are looking for (the symptoms of the disease in question, in this case) occurs quite infrequently, then few, if any, cases will be seen in either of the groups, so it would be impossible to tell whether there was any difference between them. For example, if we wanted to detect one case in 10,000 people, then we would only expect ten cases if we had 100,000 people in each group. This would be an expensive study, yet still would not result in enough cases for reliable statistical analysis.

If we are looking for the risk factors involved in a rare disease – say, one that affects only one in a million people – then a cohort study becomes effectively impossible. In such a situation, a different form of study, known as a case-control study would need to be used. Here, things are done the other way around. We start with a group of patients who already have the disease in question, and another group of people without the disease (but similar in other respects), and then assess whether or not they have previously been exposed to the risk factor we are investigating. Have they been vaccinated? Do they live near a nuclear power station? Have they been infected with this bacterium? Do they have an identifiable genetic difference, or whatever?

Whichever form of study is being used, ideally it should be done 'double-blind', that is, neither the people being investigated nor the investigators should know which group they are in. This is to avoid (or at least reduce) the likelihood of bias creeping in. If someone is asked whether they have eaten any pickled gherkins in the last six months, they might rack their brains more thoroughly if they think the investigator is trying to find out if this might have contributed to the condition they are suffering from. Similarly, the investigators might examine the test results more carefully if they know that this person is in the test group.

Clearly, it is not always possible to make the study fully blind, although various subterfuges can be adopted to try to disguise it. It should be said that the term 'bias' here does not imply any deliberate intention to deceive. It simply means that something in the way the study was designed and carried out has unintentionally distorted the data.

Whether it is a cohort study or a case-control study, the data can be analyzed in much the same way, and this involves calculating what is called the *odds ratio*. For a cohort study, this means working out the odds of getting the disease for the test group, i.e. the number who get the disease divided by the number who do not. So, if 20 people get it and ten do not, the odds of getting it are two to one on (if you're the betting type – or two if you're an epidemiologist). The same is done for the control group, and then one is divided by the other. This is the odds ratio (OR). If exposure to the supposed risk factor has no effect, then the odds will be the same for both groups, i.e. OR = 1. If the exposed group is more likely to get the disease, then the odds for that group will be higher, i.e. OR > 1.

However, even if we get an OR > 1, it does not mean that we have established an association; it could be just due to chance. It is therefore necessary to work out how likely (or unlikely) it is that this might be just chance, by calculating the *confidence intervals*. This gives a range of values within which we can be 95 per cent certain that the true value of the OR lies. If the lower end of that range is still greater than 1, then there is a statistically significant association.

It is important to realize that having established an association between the risk factor and the disease does *not* mean that this is proof that the factor in question actually causes the disease. There could be a number of other reasons for having such an association, the most important being that both the supposed risk factor and the disease in question are associated with something else. In a well-conducted study, most of the obvious factors (e.g. age, sex and a range of other possibilities) will be taken care of by matching the test and control groups, but it is always possible that there is some other factor that has not been thought of.

For a cohort study, the data can be compared in a slightly different way, which gives a more intuitive parameter known as *relative risk*. This involves a comparison of the incidence of the disease (number of cases divided by the total size of the group) for the test and control groups. As with the OR, a value of 1 indicates that there is no association at all between the supposed risk factor and the disease – an individual is just as likely to get it whether or not they have been exposed to this risk factor. Values greater than 1 suggest an association (subject to the caveats above). In this case, however, it can be put into words that are readily understood – so that if the relative risk is 2, for example, then someone is twice as likely to get this disease if they have been exposed to this risk factor. This is the sort of result that is usually reported in the media, when they say things like 'eating beetroot increases your risk of deafness by 50 per cent' (not a real example!) – it means the relative risk is 1.5.

One further important factor that often gets ignored in reports of these associations is the actual level of risk involved. Something that increases the risk of a disease by 50 per cent sounds very dramatic but, if the actual frequency of cases is taken into account, it may not be that serious. Suppose this condition affects two people in 10,000 in the absence of the risk factor. An increase of 50 per cent in the number of cases means that the risk goes up from 2 to 3 per 10,000 – hardly enough to convince us that we should adopt a major change to our lifestyle.

10.5　Superbugs and killer viruses

Terms such as 'superbug' or 'killer virus' crop up frequently in the media, with the implication that these microbes are unusually dangerous. How valid is this? Are they especially likely to infect you and make you seriously ill, or perhaps even kill you?

The terms are most commonly applied to two bacteria – MRSA (methicillin-resistant *Staphylococcus aureus*) and *Clostridium difficile* (*C. diff.*). Both of these organisms, and the diseases they cause, were described in Chapter 3, where we saw that they mainly affect patients in hospitals and other institutions such as old people's care homes. They do cause problems there, and many deaths, but it is important to remember that these are people who are highly sensitive to infection because of reduced immunity or other underlying diseases, and also that these environments are conducive to cross-infection. Neither of these bacteria are of major concern in the community. To anyone who is otherwise reasonably healthy, they are not really a cause for concern; they are a problem, but they are not as dangerous as the term 'superbug' implies.

The word 'superbug' also carries an implication that pathogenic bacteria evolve to become more virulent, which might seem to make sense in evolutionary terms if pathogenicity is regarded as the goal towards which they aspire. To examine this concept more rigorously, we need to consider where the selective advantage lies.

If we consider the simple case, where an individual catches the disease only from another person who is already infected (the source of infection), then the likelihood of infection will depend (among other things) on how many other people the infected person comes into contact with while they are infectious. If the pathogen is very dangerous, and the person infected dies quickly, they will not pass it on to many people. Even if they do not die straight away, but are very ill and stay at home in bed, they will not transmit it very well. However, if the pathogen changes so as to become *less* virulent, so the infected person neither dies nor stays at home but continues to go about their normal business, then they will transmit it to a larger number of people. This altered pathogen will, therefore have a selective advantage and will become the predominant form. The evolutionary pressures will thus operate so that a well-adapted parasite would cause few, if any, symptoms of disease but still retain infectivity. Pathogenicity is therefore a manifestation of an imperfect adaptation of the host-parasite relationship (although there are exceptions, such as cholera, where the symptoms of disease increase transmission).

Why, then, are there a number of highly dangerous pathogens, such as the virus that causes Lassa fever? The notable feature of the vast majority of extremely dangerous pathogens is that an infected person is *not* the sole source of infection. These microbes primarily infect animals, where they maintain a stable host-parasite relationship, and they only infect humans incidentally. We call such diseases *zoonoses*. Bubonic plague, which was covered in Chapter 3, is one example, where the disease circulates in the rat population and only occasionally gets transmitted to humans by rat fleas. Rabies is another example of a zoonotic disease, as it is maintained in animal populations and is rarely transmitted from animals to people. The *E. coli* EHEC strains also have their original source in animals, although, in this case, infected people are also a source of infection.

So, in general, for most pathogens, evolution will tend to operate to reduce their virulence rather than enhance it. The better we get at controlling infectious diseases, by identifying cases early and dealing with them effectively, then the stronger those pressures would appear to be. Indeed, we may have seen this happening with *Streptococcus pyogenes*, which used to be a highly feared pathogen, causing diseases ranging from severe wound infections (leading to rapidly fatal septicaemia) to scarlet fever. It is now encountered relatively rarely, and scarlet fever, in particular, is now such a mild disease (in the UK at least) that cases are rarely reported (although they do occur). The bacterium seems to have become attenuated. Although the reasons are not known, and may involve many factors, it is at least conceivable that the effectiveness of penicillin treatment may be one factor, as the bacterium has not developed resistance to this antibiotic. Of course this is not true for all pathogens, many of which have developed antibiotic resistance (see Chapter 4). The development of resistance is one situation where evolutionary pressures have clearly operated so as to increase the problem.

10.6 Microbes and climate change

We hear a lot about climate change these days, but not much about its interactions with microbes. This is an attempt to put that right. To avoid misunderstandings, let me clarify my position to start with. There are two clear facts; firstly, the climate is changing; and, secondly, we are pumping out an ever increasing amount of carbon dioxide (and other gases) which are known to have a 'greenhouse effect', reducing the rate at which heat is lost by the Earth.

To what extent is the climate change due to the greenhouse gas emissions? Climate change is not a new phenomenon. The Earth's climate has fluctuated radically over a geological time scale (think of the ice ages) and, even over the last few centuries, there are well-documented fluctuations on a lesser scale. Part of the changes that we are currently observing can be ascribed to these natural fluctuations (whatever the causes of those changes might be). However, the evidence is strong that human activity is playing a role in at least exacerbating those trends, even if we have to leave open the question of precisely how much. From that follows even more uncertainty as to what is likely to happen in the future.

I want to concentrate here on the relationship between microbial activity and climate change. What impact do microbes have on climate change, and how will climate change affect the microbes?

Microbes, especially those in the oceans, make up a substantial proportion of the Earth's biomass. Furthermore, the photosynthetic ability of these microbes – cyanobacteria and micro-algae – results in a massive conversion of atmospheric carbon dioxide into organic matter. This has been estimated at $10^9 - 10^{10}$ tonnes of carbon dioxide per year, or about one third of the total emissions of CO_2. As these organisms die, or get eaten by other organisms which, in turn, die, this organic matter sinks to the bottom of the sea and accumulates in the sediment. This has a profound effect in slowing the accumulation of carbon dioxide in the air. Thus, the effect of climate change on this system will influence predictions of what will happen to our climate.

Superficially, it might be thought that an increase in CO_2 might lead to increased photosynthesis, and a corresponding increase in carbon fixation. However, as more CO_2 dissolves in seawater, it will become more acidic. The acidity of seawater has increased by 30 per cent over the past 150 years – or, to put it another way, the pH has dropped by 0.1 (the current value is about 8.1). If CO_2 levels double, the pH of seawater is expected to fall by another 0.3–0.4 units.

What would this do to the microbial content of the oceans? We can look at a similar event that occurred about 55 million years ago, when there was a spike in atmospheric CO_2 levels, and extensive global warming (5–7 °C). Ocean sediment from that period has relatively few fossils with carbonate shells. Much of the carbonate that is formed in the oceans is due to the activity of a group of marine algae known as coccolithophores, which make a shell containing overlapping plates of calcium carbonate. As levels of CO_2 rise, there is a shift of the population in favour of species with thinner shells. The outcome is an almost linear decrease in the amount of calcium carbonate formed as carbon dioxide levels rise, so, rather than the microbes buffering the effect of increased CO_2, the converse may actually be the case.

What about the effect of temperature? If the sea gets warmer, will these microbes grow faster and fix more carbon dioxide, thus buffering the effect? As the ice caps melt, and sea level rises, there will be more surface area available for them, which will also increase the amount of carbon dioxide fixed. On the other hand, though, if the water is warmer, then the solubility of carbon dioxide will decrease, which could have the opposite effect. Also, the effect of the warmer sea on the microbes is difficult to predict. Some evidence suggests that the number of microbes tends to increase under warmer conditions, but the cells are smaller, so they might not sink to the bottom of the sea so readily.

The overall message is that the position is extremely complicated, and scientists who are attempting to predict the extent and consequences of climate change have a very difficult job.

The other side of the coin is whether climate change will cause an increase in infectious diseases, especially in temperate countries like the UK. One example that is often quoted is malaria, so it is worth considering that specifically. If the UK gets warmer, will we see an increase in mosquito populations and, hence, a spread of malaria? The historical record is quite informative. Malaria is not identified as such in these old records, but the fever known as ague is very likely to have consisted, at least in part, of what we now call malaria. From 1840–1910, there were over 8,000 deaths from ague in the UK, with the level declining steadily over that time.

The main reasons for the decline appear to be drainage of wetlands and an increase in the number of cattle. The main vector for malaria in the UK, the mosquito *Anopheles atroparvus*, is happy feeding on cattle – so, with more cattle around, the mosquito bites people less often. The wetlands have continued to decline, and the cattle numbers have continued to increase, and these trends more than outweigh the effect on mosquito populations of a 1–2.5 °C rise in temperature. To this, we can add the effect of control measures – identifying and treating cases of malaria as they arise. Over 50 years from 1953, there were some 50,000 imported cases of malaria in the UK but there was no spread of the disease to anybody else, because the control measures were adequate.

Malaria, therefore, does not seem to be a climate change-related risk – but what about other diseases that are carried by insects or other vectors? In Chapter 3, I described one example: the cattle disease bluetongue (spread by midges), which has spread northwards probably as a consequence of climate change. As another example, in 2007 in Italy, there was a local outbreak of infection with a virus called Chikungunya, originating from a recent arrival from India and spread by a species of mosquito which has established itself in Italy, presumably due to 'global warming'. There is a range of other diseases being considered as potential risks, but it is unlikely we will have a major problem, provided that suitable preparations are made.

10.7 Microbes and non-infectious diseases

This may seem an odd title. Surely the point about microbial diseases is that they *are* infectious? They are passed from one person to another – by direct contact, inhaling or ingesting infectious particles, or by means of a vector such as an insect. To this, we can add diseases (e.g. rabies) that are caught from an infected animal, and those (such as Legionnaire's disease) which we get from exposure to an environmental source. Although in the latter two examples, we do not usually catch the disease from an infected person, the general dogma remains – that microbial diseases are infectious.

As is often the case with generally accepted dogmas, an excessive adherence to it can prove damaging. It can blind us to the possibility that some apparently non-infectious diseases may nevertheless be caused by microbes, or at least have a microbe as part of the cause. Predominant amongst such diseases is cancer. We do not 'catch' cancer by contact with a cancer patient, and yet we now know that some forms of cancer do have a microbe as one factor that leads to the development of the disease. One such example, which we have already encountered, is *Helicobacter* which, as well as causing gastritis and peptic ulcers, is involved in causing certain forms of cancer of the gut (see Chapter 3). However, it is far from the sole cause; very many people carry *Helicobacter* in their gut, yet only a very small proportion of those people develop one of these cancers.

This was not actually the first example of a microbe causing cancer. The history of viruses and cancer goes back to the early years of the 20th century. In 1908, Wilhelm Ellermann and Oluf Bang found that something that could pass through a filter (which would intercept bacteria or anything bigger) was capable of causing leukaemia in chickens. This was followed three years later by the discovery, by Peyton Rous, of another virus affecting chickens – the Rous sarcoma virus (RSV). This discovery was so revolutionary, and so difficult for scientists to accept, that it was not until 1966 that Rous was awarded a Nobel Prize.

The Epstein-Barr virus (EBV), named after its discoverers Anthony Epstein and Yvonne Barr, is best known as the cause of glandular fever, but it is also part of the cause of forms of cancer in some regions of the world. These include Burkitt's lymphoma, first described by Dennis Burkitt in Uganda in 1958, which is the commonest childhood tumour in Africa, and naso-pharyngeal cancer (affecting the nose and throat) in China and Southeast Asia.

Another well-known example is the human papilloma virus (HPV), which typically infects skin and other epithelial cells. Normally, skin cells stop dividing as they differentiate, and they are eventually shed. HPV interferes with this process, so the cell keeps dividing, giving rise to benign warts and verrucas. These are not a major problem, as they do not spread. However, it only takes some changes in the host cell for this process to lead to the development of a malignant tumour. The most familiar is cervical cancer.

So there are plenty of examples of viruses causing cancer. The examples above are not an exhaustive list – it is estimated that, globally, 15–20 per cent of cancers are caused by viruses. The unusual, and exciting, feature of the *Helicobacter* story was that it was an example of a *bacterium* causing cancer. This is unlikely to be the only example, and there is speculation about several others.

I should point out that, although I started this section by saying that cancer is not infectious, this is not completely true. The examples above do not really count as infectious cancer, because the viruses (and *Helicobacter*) are quite common and we do not catch cancer solely by acquiring one of those microbes. There are, however, examples (in animals) where the cancer itself is transmissible. The best known of these affects a carnivorous marsupial, the Tasmanian devil, which is unique to the island of Tasmania. In 1996, the first cases were reported of a facial tumour in these animals, which caused death in about six months. The disease spread rapidly over most of the island, leading to one of the fastest declines in population ever recorded. It is predicted that the Tasmanian devil will become extinct within 20–30 years. The rapid spread of the disease suggested that it was infectious, but it is not caused by any microbe. It is the cancer cells themselves that spread the disease, through the animals biting each other.

There is a second example of a contagious cancer, affecting dogs, called canine transmissible venereal tumour (CTVT). This is a much older disease, first described in 1876, and it may have originated thousands of years ago. It is spread by sexual contact, and again it is the cells themselves that are infectious.

Although we have here two examples of transmissible cancers, we should not be worried that the same would happen to us. We, in common with most animals, have powerful mechanisms that reject foreign tissues, which would prevent such transmission. Tasmanian devils, on the other hand, have very little diversity in this respect, and dogs are notoriously inbred, so they are less likely to reject the cancer cells.

Extending the argument to other diseases, we can look at allergies and asthma, which happen because of an inappropriate immune response. There is plenty of anecdotal evidence to suggest a connection between microbes and these conditions. In particular, they tend to be more common in societies with high levels of hygiene – familiarly presented as the hypothesis that 'dirt is good for you'. The possible link lies in control of the immune system; the suggestion here is that exposure to certain bacteria early in life is needed for the regulation of the immune system to develop properly. One example of experimental evidence of this involves studies in mice, where animals treated with antibiotics were shown to have more food allergies than the control group. However, experimental trials in humans have given conflicting results, so the jury is still out – especially as to how one might intervene to obtain a more appropriate balance of the immune response.

One of the problems is the uncertainty over which bacteria are involved. One favoured candidate is, once again, *Helicobacter*, where the proportion of people infected has been declining, globally, over the last 40 years, during which period the incidence of asthma has been increasing (but remember the earlier discussion about the MMR vaccine – two things happening together may be just coincidence). Furthermore, fewer people are infected with *Helicobacter* in countries like the UK (<30 per cent) than in, say India (80 per cent). Several studies have shown that children with asthma are less likely to have *Helicobacter* in their stomach, but this association does not prove that the presence of *Helicobacter* actually protects against asthma. The missing factor is knowing exactly how the development of the immune response is regulated in response to *Helicobacter* and other microbes.

What about other diseases? There is an abundance of human diseases where the underlying cause is still not completely understood, so it is tempting to consider a microbial factor. This includes conditions where we know what happens but not why (e.g. diabetes). The first obvious risk factor to be considered with these diseases relates to something in the genetic make-up of those individuals – and in at least some cases, modern techniques have made considerable advances towards pinning down specific genes that may be involved. This then leaves the very broad category of 'environmental exposure', which can include exposure to infectious agents. One well-established association is the risk of cirrhosis of the liver following infection with the hepatitis B virus.

There are many other candidates, some unproven and some discarded (apart from a few diehard adherents). One (controversial) example is a possible association between Crohn's disease and *Mycobacterium paratuberculosis*. This bacterium is known to be the cause of a serious gut disorder called Johne's disease, which affects many ruminants, including cows, sheep and goats. It is marked by diarrhoea and weight loss, and pathologically by thickening of the lining of the intestine. Because these changes resemble, to some extent, those seen in Crohn's disease in humans, some investigators started to look for an involvement of *M. paratuberculosis* in Crohn's disease. There are several reports of the presence of the bacterium (or its DNA, detected by PCR) in gut tissue from Crohn's disease patients, more frequently than in control patients, but not all studies have found this. Even if it is true, it does not prove causation; for example, it could be due to the bacterium taking advantage of the altered environment provided by the damaged gut tissue.

Another favourite is cardiovascular disease. We have already seen (in Chapter 2) that bacteria in the gut can contribute to heart disease indirectly, through the production of a chemical that is associated with heart disease, and this has been confirmed by studies in mice. Numerous attempts have been made to demonstrate a more direct role for microbes, with reports of a range of viruses and bacteria, including cytomegalovirus, *Helicobacter pylori* and *Chlamydia trachomatis* being associated with coronary disease. However, none of these has been substantiated. This story has been repeated over and over, for many such diseases. In particular, the measles virus has, at one time or another, been claimed to be associated with all sorts of diseases, only for such claims to fall by the wayside with more careful studies.

The main problem in attempting to find out the cause of these conditions is that many of them are multifactorial. By this, I mean that there is no single factor that 'causes' the disease, but rather that there are a large number of factors that each result in a partial predisposition to the disease. If we just think about microbes to start with, we can ask two questions. First, is it *necessary* for the disease? For typical infectious diseases, the microbe is clearly necessary – an individual does not get the disease unless infected. Someone examining suitable specimens from a patient would expect to be able to find the microbe (if the methods used are good enough, which is not always true). However, when dealing with a condition where the microbe is only one of many factors that may be responsible, it is more difficult. It may be possible to show that the microbe is present more often in such patients than in a control group, but there is much more room for doubt.

Secondly, is it *sufficient* to cause the disease? In other words, if we can detect the presence of the microbe, then would we know, even without seeing the patient, that s/he has that disease? A good example of this is that if *Plasmodium* can be seen in a blood smear, then the patient has malaria. However, even for the infectious diseases covered in Chapter 3, this is often not true. Someone may carry *Clostridium difficile* in their intestines without any problem. There are often many other factors that are necessary for disease to develop. If we are looking for a microbe that might be one factor in causing a disease, it becomes very difficult if most people can be infected without any symptoms.

Similar considerations apply to attempts to identify any genes involved (which could also interact with microbial factors). For some diseases, we know that a change in a single gene is directly responsible. For example, cystic fibrosis arises because of specific changes in the gene coding for a protein that is involved in transport of chloride ions across the cell membrane. There are a sprinkling of other diseases that are similarly due to specific genes, but these seem to be the exception. Detailed studies of the DNA of large numbers of people have shown that, for many diseases, even where there is known to be an inherited predisposition, only a small proportion of the inherited component can be readily identified. It seems that the inheritance is due to the cumulative effect of a large number of genes, each of which has a very small, possibly undetectable, influence on its own. If we then add to this complex genetic background all the various environmental influences – including diet and stress – we can see something of the difficulty in assessing whether a specific microbe is also a contributory factor.

Even if the microbe *is* a necessary factor, it is unlikely to be both necessary and sufficient; it will only cause the disease if the other predisposing factors are present. We can see this if we look at a well-established situation, such as *Helicobacter* causing gastritis. It does seem to be a necessary factor – if *Helicobacter* is eliminated, using antibiotics, the gastritis disappears. However, it is not sufficient on its own. Many people (50 per cent worldwide) carry *Helicobacter*, but most of them do not show any symptoms of gastritis. We are thus left with the concept that some of these diseases may have a microbial factor as part of their cause, but we have not been able to identify it.

There is one further complication for some diseases. In order to have any chance of establishing the cause, we must have a consistent and unambiguous case definition. In other words, we have to be sure that all the patients do actually have the same disease. If we look again at the situation with *Helicobacter* and gastritis, the bacterium is only responsible for one type of gastritis. It is possible to differentiate various sorts of gastritis, using endoscopy to examine the tissues lining the stomach. If this were not so, it would have been much more difficult, if not impossible, to establish the role of *Helicobacter*; we would just have a lot of people with gastritis, many of whom had no *Helicobacter* present. However, for many of the conditions we would like to consider, ranging from chronic fatigue syndrome to autism, the case definition is controversial. Without a precise and unambiguous definition, it can be next to impossible to define a cause or predisposing factors.

The upshot of all this is that there are several apparently non-infectious diseases that have been shown to have the presence of a microbe as at least part of the cause. It is likely that further examples will emerge in years to come.

10.8   Epilogue

We have covered a lot of ground in this journey through the world of microbes. As in any journey, some places have been explored more thoroughly than others, and I am very conscious of the amount of information that has been left out. I am sure some of my colleagues would think I have dwelt too long on bacteria and given scant regard to some of the other microbes, especially the fungi. However, choices have had to be made, and mine have reflected my own experience and interests, while trying not to ignore totally other aspects of this fascinating world.

And there is so much still to be found out. So many unknown microbes exist, especially in the environment. Even with those that we do know about – even those we think we know quite well – there is so much still to be discovered about the ways in which their genes interact with one another, let alone how the cells work in complex communities.

There are those who argue that trying to reduce everything to a set of biochemical and molecular reactions is to lose sight of the beauty of nature. To my mind, that misses the point. To understand how, and why, a flower develops in just the way it does, adds enormously to our appreciation of its beauty. And it is just the same with microbes. The subtlety and complexity of their structures and development is beautiful in itself.

Appendix 1: Explanations

The purpose of this section is to provide rather more detailed information on some aspects of the molecular background to the material covered in this book, which may be helpful to some readers.

A1.1   Monomers and polymers

Many substances within the cell are *polymers*. These consist of chains of individual units (*monomers*). In some cases, the monomers within a polymer are all identical but, more often, the polymers are composed of a variety of similar monomers. They range from structural and storage materials, such as polysaccharides, to more complex, information-containing polymers (nucleic acids and proteins).

A1.1.1   Sugars and polysaccharides

Chemically, the term *sugars* encompasses a much wider variety of substances than just the sugar we are most familiar with, which is chemically known as sucrose. All sugars are carbohydrates, that is they share the basic formula $(CH_2O)_n$. For most of the examples we will encounter, n is either 6 (called hexose sugars, e.g. glucose) – or 5 (pentose sugars, e.g. ribose, which forms part of RNA). Sucrose is composed of two different sugars, glucose and fructose, joined together, i.e. it is a *disaccharide*. Lactose is also a disaccharide, composed of glucose and galactose. Fructose and galactose are both hexose sugars.

There are several types of sugar polymers (polysaccharides) encountered in this book. Starch and glycogen are both storage compounds, the former being

---

Understanding Microbes: An Introduction to a Small World, First Edition. Jeremy W. Dale.
© 2013 John Wiley & Sons, Ltd. Published 2013 by John Wiley & Sons, Ltd.

commonly found in plants, while the latter is restricted to animals. Both are polymers of glucose, the difference being mainly in the length of the chains and the degree of branching. Cellulose is a structural polymer in plants, and it is also composed of chains of glucose, but the glucose units are joined together in a different way from that found in starch and glycogen, This apparently simple difference has major consequences. Not only does it make cellulose stronger, but also it makes it difficult to break down. No animals can break down cellulose, as they do not have the enzymes needed to digest it. This ability to degrade cellulose is strictly a microbial activity. Animals such as ruminants, which rely on plant material for their food, use microbes in their gut to break down the cellulose.

There are many other structural polymers, with more complex composition. These include lignin (a tough polymer, difficult to degrade, that is a major component of wood), chitin (found in the cell wall of filamentous fungi), and peptidoglycan (a modified polysaccharide with peptide side chains that is the main structural component of bacterial cell walls).

In this book, we encounter bacterial capsules, which are also polysaccharides. These are protective, rather than structural, and in particular they protect some pathogenic bacteria from phagocytosis. They are also antigenic and are, therefore, important in our immune responses to infection; we have encountered some examples where they are used as vaccines.

This is not anywhere near a complete picture. There are many other polysaccharides (and related structures), some of which include sugar derivatives with other substances attached to them. However, this list covers those referred to in this book.

A1.1.2 Amino acids and proteins

Proteins perform a variety of functions. Some, for example, such as keratin in nails and hair, and collagen in tendons, are structural proteins, while actin and myosin are the contractile proteins that form the basis of our muscles. However, the proteins that we encounter most in this book are enzymes, which are the catalysts that drive all the biochemical reactions in a cell. I will deal with enzymes more fully in a subsequent section.

Proteins are polymers of amino acids. An amino acid is characterized as having an acidic (COOH) group and a basic (NH_2) group. In a protein, the amino acids are joined in a chain by links (peptide bonds) between the COOH group of one amino acid and the NH_2 group of the next. A short chain of amino acids is called a peptide, and longer ones polypeptides or proteins – but there are no absolute dividing lines between these terms.

There are 20 amino acids that occur naturally in proteins (leaving out some that are modified after the protein is made, and yet others that occur in other contexts in the cell). These proteins vary in size considerably but, if we consider a typical protein as having a sequence of 300 amino acids, with any of the 20 amino acids at each position, we can see that there is an enormous variety of possible sequences. Furthermore, interactions between the amino acids in the chain cause the protein to fold up into a complex shape, which is a major factor in determining its activity and function. An enzyme has one or more 'pockets' into which the specific substrate fits precisely, thus making the enzyme highly specific for that substrate, as well as facilitating the catalytic activity of the enzyme through interactions between the substrate and the amino acids that form the pocket.

As well as enzymes and structural proteins, there are proteins that have other functions, including control of the cell's activities. Four such activities are referred to in this book:

1. *Regulation of gene expression.* Some proteins can bind to specific sites on the DNA, so switching the expression of individual genes, or groups of genes, on or off. Other proteins (sigma factors) bind to RNA polymerase and alter its specificity, which also switches groups of genes on or off.

2. *Transport.* The cell is surrounded by a lipid membrane, and food (and other things) have to cross that barrier. Proteins within the cell membrane form channels through which selected compounds can move. In some cases, this movement is driven by the difference in concentration, so the material simply flows by diffusion; in other cases, however, the protein channel is linked to an energy-generating system which allows the material to be pumped in, which means that the cell can accumulate it even though the concentration outside is low. Similar concepts apply to the export of material. Both are important in influencing the sensitivity of the cell to antibiotics, as well as its ability to take up food and excrete waste products.

3. *Sensing the environment.* Some proteins (sensors, or receptors) in the cell membrane have the job of detecting the presence of specific chemicals in the surrounding medium. They act like enzymes in specifically recognizing, and binding to, the chemical in question. Instead of having a catalytic effect on the chemical, though, the binding alters the shape of the protein, and that change is transmitted across the membrane to the inside of the cell. Here, the protein interacts with other proteins (regulators) that affect gene expression within the cell. Some membrane receptor proteins (photoreceptors) respond to light rather than chemicals – these are related to the receptor proteins in our eyes.

4. *Movement.* Many bacteria move by means of rotation of flagella – long polymers of proteins that project from the cell surface. The rotation is carried out by other proteins in the cell membrane that act as a motor. Other proteins are responsible for movement of a different nature, for example the movement of replicated DNA towards the ends of the cell.

A1.1.3   Nucleic acids

Nucleic acids (DNA and RNA) are polymers of nucleotides. A nucleotide is composed of a pentose sugar (ribose in RNA, deoxyribose in DNA), which has a phosphate group attached at one side and one of four 'bases' at the other side. These bases are adenine (A), guanine (G), cytosine (C) and thymine (T) in DNA, while RNA has uracil (U) instead of thymine. The sugars are linked together in the backbone of the nucleic acid by the phosphate groups, while the bases (which are the part that encodes the information) stick out to the side.

RNA molecules usually have just one strand (although some viruses have double-stranded RNA), while DNA (usually) has two strands side by side. These two strands are complementary, which means they are a sort of mirror image of one another. Where there is an A in one strand, there will be a T in the other, while each G is matched with a C. This enables the DNA to be copied very accurately – the two strands separate (over a very short distance) and each is copied into a new complementary strand. If the enzyme system that does the copying makes a mistake, so that the wrong nucleotide is incorporated, that mistake can be detected (because the two strands will not pair properly) and corrected.

Another important fact about double-stranded DNA is that the two strands run in opposite directions (this derives from the orientation of the phosphate linkages between the sugars). Nucleic acids are only made in one direction so, when copying a double-stranded molecule, one strand is made from left to right, as it were, and the other from right to left. We saw this at work in the explanation of PCR in Chapter 7.

DNA is the basic genetic material of all cells (although some viruses have an RNA genome). How does the information that it contains, in the form of a sequence of G, A, T, and C, determine the characteristics of the cell? The main process involves translating this message into the sequence of all the different proteins in the cell. However, there are 20 amino acids to be coded for, and only four bases in the DNA. The secret here is that the message encoded in the DNA is read in groups of three (triplets). There are 64 possible triplets, which is more than enough to code for the 20 amino acids, many of which are coded for by more than one triplet. So, for example, where the DNA has AAA or AAG, there will be a lysine in the protein; where it is GGT, GGC, GGA or GGG, the amino acid will be glycine.

The DNA itself is not read directly for this purpose. Before proteins can be made, the DNA corresponding to an individual gene (or, in bacteria, groups of genes) is copied (by RNA polymerase) into an RNA molecule, known as messenger RNA (mRNA). This process, known as transcription, provides an important regulatory step, as the cell is able to select which genes should be expressed and which should not – depending, for example, on whether a suitable substrate is present. For example, in *E. coli*, the genes needed for the breakdown of the sugar lactose are not transcribed unless lactose is present in the growth medium.

Once the mRNA is made, it is translated into protein by *ribosomes*, which are complex structures made up of several RNA molecules and a number of proteins.

The ribosome can recognize signals in the sequence of the mRNA which determine precisely where translation should start. This is vital, as there is no punctuation to indicate which triplets should be read. If translation started in the wrong place, it would result in a completely different protein, because the triplets would be different.

A1.1.4 Fats and lipids

One further type of substance in the cell needs to be dealt with. These are the lipids (or fats, in common language). They are not strictly polymers, but it is convenient to include them here.

We can start by considering the fatty acids, which consist of a hydrocarbon chain with an acidic group (COOH) at the end. The simplest example is acetic acid (familiar as the main constituent of vinegar), which has the structure CH_3COOH. If we move down the table, we get propionic acid (C_2H_5COOH), then butyric acid (C_3H_7COOH), and so on to things like stearic acid (or octadecanoic acid, $C_{17}H_{35}COOH$). Acetic acid is water-soluble (as are propionic and butyric acids) but, as the hydrocarbon chain gets longer, the fatty acids become insoluble in water, and the larger ones are solids (technically, the hydrocarbon chains are *hydrophobic*, i.e. they do not like water). These are known as *saturated* fatty acids; *unsaturated* fatty acids (which have one or more double bonds in the hydrocarbon chain) are liquids – as you see if you compare vegetable oils with animal fats like butter. Generally, fats and oils are made up of one or more fatty acids joined to a glycerol molecule.

Although the smaller fatty acids, such as acetic and propionic acid, are important in the general metabolism of the cell, the real importance of lipids occurs in derivative forms, where they form the major component of cell membranes. In particular, a phosphate group can be added (forming a *phospholipid*) or a sugar residue may be included (*glycolipids*). The importance of these molecules derives from the fact that the phosphate and sugar are hydrophilic ('water-loving') while the hydrocarbon chains are hydrophobic. This results in a molecule in which one end 'likes' being in water while the other end does not. The consequence is that the hydrophobic ends will group together, to escape from the water, leaving hydrophilic groups such as the phosphates on the outside, interacting with the water. This produces a double layer, with the hydrophobic groups sandwiched between two hydrophilic layers, and it forms the basic structure of the membranes that surround cells of all types.

Such structures will form spontaneously in water, producing either sheets or spherical structures. Under appropriate conditions, the spherical form will enclose a small amount of water containing drugs or other chemicals present in the solution in which it is formed. These are known as *liposomes*, and they can be used to transport, into cells, drugs that would otherwise not penetrate the cell membrane effectively.

A1.2   Enzymes and catalysis

Any chemical reaction is, in principle, reversible. When the gas is lit on a cooker, methane combines with oxygen to form carbon dioxide and water, together with the liberation of heat. Therefore, it should be possible to react carbon dioxide with water to produce oxygen and methane, if sufficient energy is supplied. However, there is a further obstacle to either reaction. Methane does not spontaneously combine with oxygen; it has to be lit to get a flame. This is analogous to facing a hill when cycling. Even if the bicycle is going to end up at a lower level than where it started, it is still necessary to pedal it up the hill before it can coast down the other side. This is where catalysts come in. In effect, they lower the energy barrier that prevents the reaction from happening spontaneously – a bit like road builders making a cutting through the hill so that our cyclist doesn't have to pedal up it.

Enzymes are biological catalysts. The binding of the substrate to the protein, and the interaction of the substrate with the amino acids lining the pocket (see above), lowers the energy barrier by activating the substrate so that the reaction can proceed. After the reaction, the products are released and the enzyme is free to carry out a further reaction. As with all catalysts, the catalyst ends up unchanged.

Let's look at some of the common reactions that are catalyzed by enzymes. A concept that recurs throughout involves the nature of chemical bonds. A bond between two atoms is formed by the sharing of electrons between those atoms, so a chemical reaction involves the movement of electrons within or between substrates.

A1.2.1   Oxidation and reduction, respiration and photosynthesis

The most familiar example of an oxidation reaction is the rusting of iron, where metallic iron combines with oxygen to form iron oxide. Reduction is the reverse reaction – the removal of oxygen (in this example, the removal of oxygen from iron oxide to form metallic iron). In biochemical terms, this is too limited a view of oxidation and reduction. Within the cell, there are many reactions that are best considered as oxidation or reduction, but which do not necessarily involve oxygen at all. Instead, we use a definition that involves the addition or removal of electrons.

If we consider the oxidation of iron, this involves the removal of electrons from the iron atom and their transfer to an oxygen atom. Thus, oxidation can be defined as the removal of electrons, and reduction as the addition of electrons. In order to keep everything balanced, so there are no free electrons swimming around, oxidation of one chemical is matched by reduction of another.

Biochemically, many oxidation/reduction reactions involve one of two related compounds, called NAD and NADP (their full names are nicotinamide adenine dinucleotide and nicotinamide adenine dinucleotide phosphate). For simplicity, I will just consider NAD. Some enzymes use one and some use the other, but they both work in the same way.

NAD functions as an electron acceptor. When a substrate is oxidized, the enzyme withdraws electrons from it and transfers them to NAD, which also acquires a hydrogen ion at the same time, so we then call it NADH. Thus, the substrate is oxidized and NAD is reduced to NADH. Conversely, NADH acts as a reducing agent, as an appropriate enzyme can withdraw electrons from it and transfer them to the substrate to be reduced.

We can see the importance of this concept if we consider the oxidation of a 'fuel', such as glucose, by means of the process known as aerobic respiration. Overall, the reaction is represented as 'glucose plus oxygen goes to CO_2 plus water', with the liberation of energy. In order to capture this energy, the reaction involves many stages, especially (for the purpose of this section) the extraction of electrons from glucose (and its breakdown intermediates), and their passage through a chain of cytochromes until eventually they are passed on to an oxygen molecule, which is the final electron acceptor.

It is necessary to appreciate that chemicals (especially, in this context, various cytochromes) differ in their ability to accept or donate electrons, so that there is a 'pecking order'. By sequential transfer of electrons through the chain, the energy release is controlled carefully so that it can be harnessed. This is achieved through the coupling of electron transfer between cytochromes to the production of ATP, which is the main substance used within the cell for reactions that require the provision of energy. Some bacteria can use a final electron acceptor other than oxygen (e.g. nitrate or sulphate), so enabling them to carry out respiration in the absence of air (anaerobic respiration).

Fermentation, in the strict sense, is different in that there is no electron transport chain. Instead, electrons are merely transferred from one intermediate to another within the cell. For example, in alcoholic fermentation, electrons are transferred to acetaldehyde, converting it to ethanol. Since oxygen is not involved in this process, it can be carried out in the absence of air. Although some of these reactions are coupled to the generation of ATP, the overall process is much less efficient than respiration; fermentation generates fewer ATP molecules per molecule of glucose than is achieved through the use of the respiratory chain of cytochromes (microbiologists often use the word 'fermentation' in a more casual way, to describe the ability to metabolize a substrate, e.g. 'fermentation of lactose by *E. coli*', even though the process may actually involve respiration).

We can also relate the movement of electrons to the processes of photosynthesis and carbon fixation. As in green plants, photosynthetic cyanobacteria are able to fix carbon dioxide – that is, they can use carbon dioxide and convert it into organic compounds within the cell. This does not provide them with energy; indeed, it requires energy for it to happen. The energy is provided by sunlight, which is captured by the pigment chlorophyll (some bacteria use other pigments instead, but the general concept is similar).

When chlorophyll absorbs light, it is converted into an active form which releases an electron. In one system, this electron is simply passed through a chain of intermediates (as with the respiratory chain), coupled to the production of ATP, and is eventually passed back to the non-activated form of chlorophyll, which is then ready to be stimulated again by light. Some bacteria just do this, so they get energy from sunlight but without fixing carbon dioxide or producing oxygen. However, cyanobacteria do fix carbon dioxide, i.e. they convert it into organic matter, coupled with the release of oxygen. This involves extracting electrons from water, using light-excited chlorophyll, which releases oxygen. The fixation of carbon dioxide occurs via a separate reaction, using both the ATP generated by photosynthesis and also the reducing agent NADPH (also generated by photosynthesis).

A1.2.2 Hydrolysis

Many reactions in the cell in which a substance is split into two bits occur by a process known as *hydrolysis*, which uses a water molecule to achieve the break. For example, the sugar lactose, which is a disaccharide composed of glucose and galactose, is split in this way by the enzyme β-galactosidase. In a simplified chemical form, this reaction can be shown as:

$$-C-O-C- + H_2O \rightarrow -C-OH + HO-C-$$

(in this case, it is simpler not to consider the movement of electrons, but that *is* what is involved).

Similarly, proteolytic enzymes, which degrade proteins by breaking the peptide bonds between amino acids, act via a hydrolytic mechanism, as does the enzyme β-lactamase (penicillinase), which confers resistance to penicillins by breaking a key bond in the penicillin molecule.

A1.2.3 Polymerization

Enzymes are, of course, also involved in synthesis as well as degradation, and one special example, polymerization, is worth looking at briefly. Earlier, we looked at polymers. These are, in general, made by specific polymerases, which add monomers to the end of the growing chain. The process requires energy, which is usually provided by ATP, but generally only indirectly. A different enzyme is required first, to energize the substrate monomer by adding a phosphate group (using ATP). The polymerase can then use the energy contained in the activated monomer to form the new bond with the growing polymer.

For the simpler polymers, such as the polysaccharides, this is more or less all that is needed. However, for the synthesis of nucleic acids, an extra level of specificity is needed, and this is provided by an existing strand acting as a template. The enzyme therefore selects the correct nucleotide to be added, based on that which is present in the template at that point. The polymerase is able to recognize the correct base pairing (A with T, G with C) and will only make a new bond if the base pairing is correct.

Protein synthesis is even more complex. As described earlier, it requires ribosomes, and it is worth noting that the catalytic activity involved in making a new peptide bond (between the growing chain and the new amino acid) requires activity of the RNA within the ribosome as well as the proteins. Thus, there are exceptions to the general rule that enzymes are proteins. In some cases, RNA has enzymic activity, which is a concept that may have relevance to the origin of 'life'.

Appendix 2: Abbreviations and Terminology

A2.1   Abbreviations and jargon

These are some of the abbreviations and terms used. Others are described more fully in Appendix 1, or are defined/explained as and when they are used.

A	Adenine
Acidic	Tends to shed a hydrogen ion in solution, thus lowering the pH.
ATP	Adenosine triphosphate
B cell	Antibody-producing cell
Bacillus	Used informally for any rod-shaped bacterium. Also refers more formally to a specific bacterial genus, in which use it is italicized, as *Bacillus*.
Base	In general, any chemical that is *basic*. In this book, it refers specifically to the nucleic acid constituents A, G, C, T, U.
Basic	Tends to acquire a hydrogen ion in solution, thus raising the pH.
BCG	Bacille Calmette-Guerin
C	Cytosine *or* Carbon
CO_2	Carbon dioxide
Coccus	A round (spherical) bacterium
DNA	Deoxyribonucleic acid

Understanding Microbes: An Introduction to a Small World, First Edition. Jeremy W. Dale.
© 2013 John Wiley & Sons, Ltd. Published 2013 by John Wiley & Sons, Ltd.

E. coli	*Escherichia coli*. Note that it is the usual practice, when using bacterial names, to give the full name at first mention, then subsequently to abbreviate the genus name (the first part) to a single letter or, sometimes, to a longer abbreviation if there is a risk of ambiguity. I have used full names more often than that, for clarity.
Electron microscope	Uses a beam of electrons (rather than light) to form an image. The only way that it is possible to 'see' viruses. It is also used for looking at detailed structures of bacteria and other things.
Eukaryote	An organism with a nucleus and other intracellular structures. Includes algae, protozoa and fungi (as well as plants and animals).
G	Guanine
Hydrophilic	'Water-loving'. Hydrophilic chemicals are water-soluble. Similarly, hydrophilic groups in more complex molecules will be attracted to the water surrounding them.
Hydrophobic	'Water-hating'; the opposite of *hydrophilic*.
Hypha	Filamentous structure in many fungi and actinomycetes.
mRNA	Messenger RNA
MRSA	Methicillin-resistant *Staphylococcus aureus*
Mycelium	Collection of hyphae
N	Nitrogen
NAD	Nicotinamide adenine dinucleotide
NADP	Nicotinamide adenine dinucleotide phosphate
PCR	Polymerase chain reaction
pH	A measure of the degree of acidity; low pH means more acidic. To put it formally, the pH is the negative logarithm of the hydrogen ion concentration, and therefore a 1 unit change in pH means a tenfold change in hydrogen ion concentration.
Prokaryote	An organism lacking a nucleus. Includes bacteria and Archaea.
Ribosome	Structure containing rRNA and a number of proteins; carries out translation.
RNA	Ribonucleic acid
rRNA	Ribosomal RNA
SRB	Sulphate-reducing bacteria
T	Thymine
T cell	Cell involved in cellular immunity.
Transcription	Synthesis of RNA using a DNA template.
Translation	Protein synthesis, using coding information in mRNA.
tRNA	Transfer RNA; the type of RNA involved in protein synthesis.
U	Uracil

A2.2   Numbers

In microbiology, we are often dealing with very large numbers of microbes. It is very cumbersome to refer to a thousand million bacteria as 1,000,000,000 (and sometimes we get numbers much bigger than that). Instead, we use what is called 'scientific notation'. For this, we use the number of times that 10 is multiplied by itself to get the number required. So, for example, one thousand is three tens multiplied together ($10 \times 10 \times 10$), which is 10^3, and a million is six tens multiplied together $= 10^6$. This can also be thought of this as 1 with 6 zeros after it. So:

$$10^2 = \text{a hundred or } 100$$
$$10^3 = \text{a thousand or } 1,000$$
$$10^6 = \text{a million or } 1,000,000$$
$$10^9 = \text{a thousand million (a billion) or } 1,000,000,000$$
$$10^{12} = 1,000,000,000,000$$
$$\text{and so on.}$$

Multiplying and dividing these large numbers then becomes very easy. To multiply 100 by 1,000, which is $10^2 \times 10^3$, simply add the indices together to get 10^5 (or 100,000). For division, subtract the indices – so 1,000 divided by 100 ($10^3 \div 10^2$) is 10^1 or just 10.

We can easily extend the system to cope with numbers that are not exact multiples of 10. If we have 300 bacteria, this is 3×100 or 3×10^2. So, for example, if we have a bacterial culture that has 3,400 cells per ml and we want to know how many there are in one litre (which is 1,000 ml), then we can write the concentration as 3.4×10^3 cells per ml, and multiply that by 1,000 (10^3) to give the answer as $3.4 \times 10^3 \times 10^3 = 3.4 \times 10^6$.

We also come across very small numbers, much less than one, which can be dealt with in a similar way, although perhaps less obvious. The key to it is that mathematically we can write $\frac{1}{10}$ as 10^{-1}. From the rules above about multiplication and division, we can see that one thousandth ($\frac{1}{1000}$), which is $\frac{1}{10} \times \frac{1}{10} \times \frac{1}{10}$, or ($10^{-1} \times 10^{-1} \times 10^{-1}$), is 10^{-3}. More generally, we can say that $10^{-n} = \frac{1}{10}^n$. So:

$$10^{-1} = \frac{1}{10^1} = \text{one tenth, or } 0.1$$

$$10^{-2} = \frac{1}{10^2} = \text{one hundredth, or } 0.01$$

$$10^{-3} = \frac{1}{10^3} = \text{one thousandth, or } 0.001$$

$$10^{-6} = \frac{1}{10^6} = \text{one millionth, or } 0.000001$$

A2.3   Units

The basic units of length, mass, and volume are, respectively the metre (m), gram (g) and litre (l). These are all uncomfortably large when dealing with microbes, so we use smaller units that are obtained by dividing by multiples of a thousand, and are accompanied by prefixes to indicate this:

$$m = milli = \frac{1}{1000} \text{ or } 10^{-3}$$
$$\mu = micro = 10^{-6}$$
$$n = nano = 10^{-9}$$
$$p = pico = 10^{-12}$$
$$f = femto = 10^{-15}$$

So the size of a bacterium would usually be referred to in μm (micrometre) dimensions, while a virus would be in the nm (nanometre) range.

Appendix 3: Further Reading

I have (deliberately) not cluttered up the text with references. These days, web searches can find the sources of information for you. Therefore, this list just indicates the most useful books that I have on my shelf.

Introductory level

Dixon, B. (2009). *Animalcules: The Activities, Impacts, and Investigators of Microbes*. ASM Press.

Maczulak, A. (2011). *Allies and Enemies: How the World Depends on Bacteria*. FT Press (Pearson Education).

Money, N.P. (2011). *Mushroom*. Oxford University Press.

Wassenaar, T.M. (2012). *Bacteria: The Benign, the Bad, and the Beautiful*. Wiley-Blackwell.

Special topics

Dale, J.W. & Park, S.F. (2010). *Molecular Genetics of Bacteria*. 5th edition. Wiley-Blackwell.

Dartnell, L. (2007). *Life in the Universe: A Beginner's Guide*. Oneworld Publications.

Rosen, W. (2008). *Justinian's Flea: Plague, Empire and the Birth of Europe*. Pimlico.

Sherman, I.W. (2006). *The Power of Plagues*. ASM Press.

Varnam, A.H. & Evans, M.G. (2000). *Environmental Microbiology*. Manson Publishing.

More serious reading

Adams, M.R. & Moss, M.O. (2008). *Food Microbiology*. 3rd edition. The Royal Society of Chemistry.

Black, J.G. (2012). *Microbiology: Principles and Explorations*. 8th edition. Wiley-Blackwell.

Collier, L., Kellam, P. & Oxford, J. (2011). *Human Virology*, 4th edition. Oxford University Press.

Scragg, A. (2005). *Environmental Biotechnology*. Oxford University Press.

Understanding Microbes: An Introduction to a Small World, First Edition. Jeremy Dale.
© 2013 John Wiley & Sons, Ltd. Published 2013 by John Wiley & Sons, Ltd.

Subject Index

Acetic acid 89, 205
Acetobacter 89
Acidic environments 117–18, 163–5, 194
Acidithiobacillus 163
Aflatoxin 86
Agar 7–8
Agrobacterium tumefaciens 111
AIDS 52–3, 76, 79, 81, 95, *see also* HIV.
Alcohol *see* Ethanol
Algae 8–11, 116, 162, 169, 194
    blooms 11, 101, 103–4, 105
    food 90
    lichens 114
    photosynthesis 10, 98, 100, 102, 194
Alkaline environments 118
Allergies 55, 197
Amino acids 85, 202
    biosynthesis 160
    diet 25, 160
    protein degradation 85, 208
    protein formation 14, 16, 127, 204–5, 209
    protein structure 75, 179, 202–3, 206
Amoebae 10, *see also* Protozoa
Ampicillin 124, 171–2
Amylase 20
Anaerobes 20–1, 102, 162
Anaerobic digestion 113, 162, 169–70

Anaerobic environments 14, 15, 102, 117, 148
Animal cells 5, 8, 9–10
Animal diseases 58–62, 152, 193
Animals
    antibiotics 78
    germ-free 24
    normal flora 12, 24–6
    zoonoses 42, 193
Anthrax 152
Antibiotic resistance
    mechanisms 74–5, 123–4, 126, 171–2
    misuse of drugs 77–8
    mutants 45, 53, 75–6, 81, 121–2
    plasmids 75–7, 123–4
Antibiotics *see also* individual antibiotics
    action 16, 74–5
    antiprotozoal drugs 79–80
    antivirals 76, 80–1
    discovery 72–3, 170–2
    effect on normal flora 24, 51
    limitations 52, 77–9, 148
    production 159–60, 171–2
    selective toxicity 45, 74–5, 79, 80
    sources 73, 77, 124, 159, 171–2
    treatment 38–40, 50–1, 73–6, 171–2
    use in animals 78
Antibodies *see* Immunity

Understanding Microbes: An Introduction to a Small World, First Edition. Jeremy W. Dale.
© 2013 John Wiley & Sons, Ltd. Published 2013 by John Wiley & Sons, Ltd.

Index of Names

Understanding Microbes: An Introduction to a Small World, First Edition. Jeremy W. Dale.
© 2013 John Wiley & Sons, Ltd. Published 2013 by John Wiley & Sons, Ltd.